Augustin Planque, Mary McMahon

Fetichism and Fetich Worshipers

Augustin Planque, Mary McMahon

Fetichism and Fetich Worshipers

ISBN/EAN: 9783337037000

Printed in Europe, USA, Canada, Australia, Japan

Cover: Foto ©berggeist007 / pixelio.de

More available books at **www.hansebooks.com**

FETICHISM
AND
FETICH WORSHIPERS.

By
Rev. P. BAUDIN,
Missionary on the Slave Coast of Africa.

A HUMAN SACRIFICE TO UGUN, THE GOD OF WAR (SLAVE COAST).

Sold for the benefit of the Society of African Missions, (Lyons) France.

TRANSLATED BY MISS M. McMAHON.

NEW YORK, CINCINNATI, AND ST. LOUIS:
BENZIGER BROTHERS,
PRINTERS TO THE HOLY APOSTOLIC SEE.
1885.

PREFACE.

THE painful but necessary task which has been imposed upon us of appealing to the generous Catholics of this country in behalf of the Society of the African Missions has decided us to offer to the public the English translation of the book "Fetichism and Fetich-Worshippers," which is an account of twelve years' experience among these people. It is an interesting study of the worship of countless black tribes who inhabit Equatorial Africa. The various rites and ceremonies, including human sacrifices, still prevailing among the blacks are fully described, showing how human nature is degraded in these countries by a singular mixture of materialism and spiritualism. It treats also in a graphic manner of the trials and triumphs of the Catholic Church in these fever-breeding climates, where the zealous priest finds his tomb after three or four years of missionary labor, and often in a shorter time. We are convinced that the book will be well received, from the fact that it has already attracted the attention of the French clergy because of its connection with the study of the existence of God, in which respect it may be considered as a corollary of the study of theology. We deem it to the purpose to add to this study of Fetichism an account of the work already accomplished in the missions confided to the care of this Society, such as the Slave Coast, the Gold Coast, the Kingdom of Dahomey, where every year hundreds of human victims are sacrificed to false divinities, along the Niger, and in Egypt, which country has lately been so ravaged by the terrible scourges of war, pestilence, and persecution.

Christian charity requiring us to share with our fellow-man the goods which the bounty of God has bestowed upon us, we trust that the prayers and contributions of the reader will come to our assistance.

REV. F. MERLINI, *African Missionary*.

HOUSE OF THE IMMACULATE VIRGIN,
    Lafayette Place,
        New York City, N. Y. (P. O. Box 3512.)

PART FIRST.

FETICHISM;

OR,

THE RELIGION OF THE NEGROES OF GUINEA.

IN the midst of the explorations and scientific expeditions which have gradually robbed Africa of her mysteries, *fetichism* has guarded its own. Up to the present time, this word has awakened in Europe only a very vague idea of the adoration of animal matter, and a sentiment of profound pity for the unfortunate black fetich-worshippers. We grant that appearances favor these sentiments. The European on arriving in Guinea encounters at every step in the negro villages idols of wood or clay, as grotesque as they are unclean, rudely made, and daubed with cock's blood and palm-oil by their stupid adorers. One glance suffices to fill the European with contempt for this worship; but when he soon learns that these shapeless divinities thirst for human blood, and that human victims are immolated to appease them, immediately adding indignation to contempt he execrates fetiches and fetich-worshippers, considering them thereafter unworthy his attention. This explains the imperfect, even false idea that prevails concerning fetichism. What is called fetichism is but the material covering; but if by the aid of profound study we are enabled to read through this veil, fetichism appears very different, and we are astonished to discover under this coarse and repellent exterior a chain of doctrines and a complete religious system, of which spiritualism forms the greater part. And what is quite remarkable, these doctrines offer striking analogies to the paganism of the civilized nations of antiquity. Replace these rudely-fashioned statues by the masterpieces of Greek art, these poor fetich-huts by Roman and Athenian temples, and under these different forms, but with identically the same attributes, fetichism will call up before the mind Neptune, Mars, Mercury, Vulcan, Æsculapius, Apollo, and other gods and demi-gods or genii which are met with in the study of pagan antiquities.

Whilst savants take such infinite pains in the study of these ancient worships, in deciphering from hieroglyphics the mysteries of bull Apis, or to discover at great expense beneath ruins some vestige of the forgotten divinities of the Babylonians, it seems as though it might not be uninteresting to explore the mysteries of fetichism which is really so near to us, and which is also the religion of millions of human beings. Fetichism in fact shares with Mahometanism all Equatorial Africa. It is the religion of the innumerable black tribes which inhabit Guinea, the Gold Coast, Ashanti, Slave Coast, Dahomey, Yorouba, Benin, and the shores of the Niger and Benue. On the north it extends to Timbuctoo, on the east to Lake Tchad, and on the south as far as Gabon and to the Congo. Among these black tribes the religious and political systems, the ceremonies of worship, and the domestic customs are so intimately connected one with the other that a knowledge of their religion is indispensable to the understanding of their history and their national organization, and above all to the effectual work of their evangelization.

Their traditions and religious doctrines suggest a people more civilized than the blacks of Guinea of the present day. And on the other hand, many customs, usages, and industries show clearly that they are a people in decadence. The wars, particularly the civil wars, which have laid waste, and still continue to lay waste, these countries, have caused them to lose what they had preserved of this ancient civilization, which was in great part Egyptian, as indicated by many customs and usages.

Their mythological system, at present very incomplete, contains many vague points and many small details difficult to reconcile; but the essential points are generally pretty well fixed in the minds of the blacks by the chants, the usages, the figures, and the symbols of the divinities placed in their temples, their own dwellings, and even engraved on the doors and columns of the houses of the chiefs and on the fetich-huts.

Therefore, though scattered over an immense extent of country, these fetich-worshippers have a certain uniformity of religious belief; their divinities are identical, differing only in name; and the particular details which we give of the blacks of the Slave Coast of Yorouba, Dahomey, Benin, and other neighboring kingdoms apply to all the fetich-worshipping nations.

FETICHISM.

(The Religious System of the Negroes of Guinea.)

The religion of the blacks is an odd mixture of monotheism, polytheism, and idolatry. In these religious systems the idea of a god is fundamental; they believe in the existence of a supreme primordial being, the lord of the universe, which is his work. Monotheism recognizes at the same time numbers of inferior gods and subordinate goddesses. Each element has its divinity, who is as it were incorporated in it, who animates and governs it, and is the object of adoration. Everything in nature forms a vast scale of deification. After the gods and goddesses there are infinite numbers of good and evil genii; then comes the worship of heroes and great men who were distinguished during their lives. The blacks also worship the dead, and believe in metempsychosis, or the migration of souls into other bodies. They believe in the existence of an Olympus, where dwell the gods and celebrated men who have become fetiches, and in an inferior world, the sojourn of the dead, and finally in a state of punishment for great criminals. They have also their metamorphosis, their sacred animals, their temples and their idols, etc. In a word, their religion is similar in all things to the old polytheism of the ancients; and notwithstanding the abundant testimony of the existence of God, it is practically only a vast pantheism—a participation of all the elements of the divine nature, which is as it were diffused throughout them all.

The following is a more detailed idea of this system.

COSMOGONY AND THEOGONY.

The Idea of God.—Although deeply imbued with polytheism, the blacks have not lost the idea of the true God; yet their idea of Him is very confused and obscure. Among the numerous gods and goddesses of the black pantheon, where all the divinities are allied by androgynism, or else in divine couples, God alone escapes both androgynism and conjugal association; nor have the blacks any statue or symbol to represent Him. He is considered the supreme primordial being, the author and the father of the gods and genii; they call Him *Olorun* (*Ol-i-orun*), that is, the possessor, and master of the negro heaven or Olympus, the abode of the gods and goddesses. A city of the Yorouba bears the name of *Bi-Olorun pelu:* "if God is with us." He is also called *Olodumaré*, "The All-Powerful;" *Oga-ogo*, "The Most Glorious;" *Elemi*, "He who possesses the breath of life;" *Emi*, "The master of human souls." This name *Elemi* is incommunicable; it belongs to God alone. The gods, say the blacks, can, like Obatala, make bodies, but they cannot animate them; God reserves this power for Himself.

However, notwithstanding all these notions, the idea they have of God is most unworthy of His Divine Majesty. They represent that God after having commenced the organization of the world charged Obatala with the completion and government of it, retired and entered into an eternal rest, occupying Himself only with His own happiness; too great to interest Himself in the affairs of this world, He remains like a negro king, in a sleep of idleness.

Thus the blacks render no worship whatever to God, completely neglecting Him, to occupy themselves with the gods and goddesses and the spirits to whom they believe themselves indebted for their birth, and their fate in this life and the next. However, although they seem to expect nothing from God, the negroes by instinct naturally address themselves to Him, and invoke Him in sudden danger or in great affliction. When they are victims of injustice, they take God to witness their innocence.

Olorun ri mi: "God sees me."

Olorun mo pé emi kó puro: "God knows I do not lie."

Olorun gbá mi o: "O God, save me."

They also swear by God, and very often in these simple words:

Olorun, Olorun! "God, God!" at the same time raising their hands to heaven.

In their salutations and conversations the name of God is frequently heard, as in the morning salutation:

O ji re ô? "Have you risen well?" to which they generally reply:

A yin Olorun! "God be praised!"

For the evening salutation they often use these words:

K'Olorun k'o cho gbogbo wa! "Oh! may God protect us all!"

But if the blacks forget God, they never cease invoking the fetiches, to whom they address themselves in all the circumstances of life, directly and not as mediators or intermediators between God and man. The gods of the first rank are sovereign masters in their domain, acting according to their pleasure and their own character, and effecting immediately evil or good. The secondary gods and the genii are often subject to the superior gods, but they have also their own domain, where they are considered free to act according to their own will.

The fetiches, from the Portuguese word *feitiço*, charm or enchantment, are called in the Nago dialect *oricha*, a word which signifies custom, religious ceremony, usage. The Oricha are divided into three very distinct classes:

1. The superior gods and goddesses.
2. The inferior gods and goddesses.
3. The good and evil genii.

I. SUPERIOR GODS AND GODDESSES.

Three principal gods occupy the first place in the negro pantheon; these are Obatala, Odudua, and Ifa.

1. *Obatala.*

The first of the superior gods is *Obatala* (*Oba-ti-ala*, " The king of whiteness and light"). White is the color consecrated to him, and a white pennant floats above his temples, which are always white. His statues, his symbols, and the insignia worn by his adorers are also white.

He is also called *Oricha nla*, " The great Oricha ;" *alamorere*, " He who possesses good earth,"—for the blacks say that it is he who forms the human body in the maternal womb, and negresses address themselves to him for the happiness of becoming mothers. This belief makes them look upon deformed persons, and particularly albinos, rather common on the coast of Guinea, as the work of Obatala, who thus creates these beings to keep his office of demi-creator in the minds of the people.

He is called, besides, *Oricha kpokpo*, " The protector of the gates of the city ;" in which capacity he is represented on horseback, armed with a lance; also *Alabalache*, " The oracle that predicts the future ;" *Oricha oginia*, " The fetich who enters man": under this name he is celebrated among the Nagos.

At Porto-Novo he is still better known under the name of *Ousé*. In all doubtful cases the king has recourse to him to discover the innocence or guilt of the accused. This fetich consists of a large hollow wooden cylinder about three feet and a half long, and as large as a man, one end of which is closed with snail-shells and the other with cloth. The fetich is placed on the head of the accused, who kneels and holds it with both hands and with all his strength. If the fetich fall forward, the accused is declared innocent; if it fall backward, he is pronounced guilty. Strange to say, notwithstanding the efforts of the accused, the fetich executes all the movements commanded by the fetich-priest: it sways from one side to the other and finally falls, forward or backward when the fetich-priest commands him to declare the guilt or innocence of the accused. Many people think that a child is put in the cylinder, and that when the child has served several years he is killed and replaced by another. The secret is thus preserved, and the king can easily and without contest render justice. When Ousé has once spoken, the case is irrevocably decided.

Obatala is the greatest of the gods; he is the first of the beings that God, the supreme Being, produced in the beginning. He placed him with other spirits in the superior region of the universe and united him to *Odudua*, who became his spouse.

2. *Odudua.*

Odudua, the great goddess of the blacks, the mother of the gods, seems to be considered as never having been created, but as eternal and coexistent with God. Odudua, who is also called *Iya Agba*, "The mother who receives," dwells in the inferior regions of the universe.

Obatala and Odudua were in the beginning tightly compressed and enclosed in a bottle-gourd, Obatala in the lid and Odudua in the bottom of the bottle, engulfed in the waters, enveloped in profound darkness, fear and hunger pervading every sense; they were only a restless mass, without form or figure, and blind. The fetich-priests tell the people that Odudua was made blind and ugly in consequence of a domestic quarrel in which Obatala tore out his companion's eyes to compel her to keep quiet. She in her anger cursed him, and said, "Thou shalt have snails for thy food." *Olorun olodoumaré* ("All-Powerful God"), be sought by Odudua to restore her sight, declared that in punishment she should remain blind, but that Obatala, for having yielded to anger, should eat snails: and in fact this is the principal sacrifice offered by the blacks to Obatala.

Obatala represents all superior things, Odudua all that is inferior; Obatala is mind, Odudua is matter; Obatala is the firmament, Odudua is the earth,—all of which is symbolized by a whitened gourd provided with a cover, which is placed in the temples.

Obatala and Odudua, the fetich-priests say, are one and the same divinity, an hermaphrodite divinity. This idea is represented by a statue which has but one foot and one arm, and a tail ending in a ball or globe.

Obatala and Odudua are met with again under the sirname of Aroni and Aja, only then they have fallen gradually from the rank of superior gods to the third rank of genii or hobgoblins.

In the temples of the more modern cities Obatala and Odudua are completely divested of their hermaphrodite characteristics, are divided into two perfectly distinct divinities, and are then represented separately: Obatala under the form of a warrior, and Odudua as a female nursing her infant. However, in the irrespective temples the symbolic gourd is generally placed in front of these statues, to recall the ancient doctrine.

In their houses, those who cannot procure a statue content themselves with the gourd, before which they offer their presents and sacrifices. Still more recently the god Obatala and the goddess Odudua have become much more separate and distinct, and are not even associated conjugally. They both inspire equal confidence, take the same rank, and are on the same footing in the receipt of honors; their personality is completely distinct, and each one has a special temple. Odudua especially, rising always in power in the minds of the people, has almost entirely lost her character of spouse and has become independent; she reigns as sovereign goddess of Ado, a modern city situated not far from Badagry.

THE GODDESS ODUDUA, AND FETISH TEMPLE AT PORTO NOVO.

The legend relates that a hunter met Odudua one day walking in the forest. The goddess proposed to live with him, and they dwelt together for a long time, given over to the pleasures of the chase and fishing, and passing the rest of the time in a cabin built of boughs at the foot of a tree in the forest. Finally the goddess, disgusted with a mortal as she had been with an immortal, departed, promising that she would always protect him and all those who would establish themselves in this place and erect to her a temple in place of the cabin. Many persons came and established themselves here, and thus was founded Ado, which means prostitution, in memory of the goddess.

The temple erected in this city is celebrated among the blacks; the neighboring kings offer an ox to the goddess on her feast-day, and, in accordance with the legend, impure games are celebrated in her honor.

These two gods Obatala and Odudua have other names and symbols which for the people constitute different gods. This multiplicity of gods is more seeming than real, but it is very lucrative for the fetich-priests.

DEPENDENT DIVINITIES OF OBATALA AND ODUDUA.

We have already said that God having created the first of beings, Obatala, He united him to Odudua. Shortly after their union Odudua gave birth to *Aganju*, a name which signifies "The desert," and to *Iyemoja*, "The mother of fish." Iyemoja had by her brother a son *Orungan* ("The middle of the day," "the air," "the firmament ").

Outraged by her son Orungan, Iyemoja, inconsolable, fled, refusing to listen to the guilty one, who pursued her in her flight, begging her to return. As he was about to overtake her, Iyemoja fell backward; her two breasts, swollen immeasurably, changed into two streams which formed a lagoon called Odo Iyemoja, the lagoon of Iyemoja near d'Okiadan. Her body becoming enormous burst open. The place is shown at Ifé the holy city of

Yorouba (Ifé signifies enlargement). From Ifé, that is, from the gaping body of Iyemoja, came forth in great confusion all the gods and goddesses, of which the principle are

Olokun.

The god of the sea and of the ocean, the negro Neptune, dwells in an immense palace under the sea. Seven enormous chains now hold him captive. In a moment of anger he attempted to destroy mankind because of their propensity to lie. He had almost exterminated them when Obatala interfered and forced him back to the sea, where he remains chained in his palace forever. From time to time his efforts to break his chains create the storms on the ocean. Animals are sacrificed to him, and sometimes human beings.

Olosa.

His wife is *Olosa* (the lagoon), who also has her palace under the waters. The crocodile is sacred to her, and is supposed to be her messenger. Sacrifice is offered to Olosa in small temples on the lagoon; sometimes they also immolate to her human victims to make her favorable to the fisheries.

But sacrifices are more frequently offered to her messenger the crocodile, who is supposed to carry to his mistress the offerings of the faithful. To this end the fetich-priests hold up for the adoration of the people the monster who is invested with this charge by the goddess. When the crocodile having the necessary marks is seen, a little cabin is made for him, or rather a few pickets with some palm-branches designate the place chosen for his dwelling, and every five days the fetich priests and priestesses bring him food.

At Porto-Novo, near the mission, there is one which is very tame. As soon as he hears the women coming singing and dancing he comes out of the water and runs to meet them. The worshippers, remaining at a respectful distance, throw him their offerings—a hen, some acasas, etc. Near the water is his temple, or rather an enclosure made of bamboo and palm-branches. There on his feast-day they dance and amuse themselves; the monster remaining near by under the water, showing his nose from time to time to see if the sacrifice is nearly ready: for this day is an occasion to him of great feasting, and his adorers may enjoy the presence of their god without fear of being devoured by him. These divinities are not perfectly secure from all misadventure, as the following fact testifies:

A Mahometan negro, rather unscrupulous about his religion, having drowned his reason in a gourd of rum, heard the chants of the pagan women on their way to the lagoon to offer sacrifice to the crocodile. Remembering how delicious was the flesh of this animal, which he had eaten in his own country,* and without reflecting to what danger his audacity exposed him, he armed himself with a large harpoon and ran to the lagoon, following the women who were singing the praises of the crocodile fetich. The monster advanced to the shore to receive its customary offering.

The Haoussa darted like an arrow through the women, crying, *Allah kbar* ("God is great"), and sprung into a canoe. The adorers of the aquatic divinity, divining his intention, clasped their hands above their heads and cried in consternation, *Ye! Ye! Oricha, ô, ma kpa ô. Ye! Ye!* ("It is the fetich; do not kill him.") But the imprudent fellow, deaf to their cries, and intent on his enterprise, stood erect in his canoe, harpoon in hand, ready to plunge it into the monster, who advanced on the surface of the water, opening his jaws to catch his accustomed feast. In an instant the Haoussa thrust forward his dart and buried it through the scales in the back of the crocodile, but losing his balance at the same time, he fell into the water and disappeared.

---

*The Haousses eat the flesh of the crocodile and consider it excellent.

The women gave a cry of terror and stood as if petrified, their eyes fixed upon the lagoon. The water bubbled up and soon became red with blood. The monster, seizing his adversary, tore and mutilated him frightfully. All that remained of the Haoussa was an indescribable human wreck, bloody fragments of which floated near the shore and were carried along by the current.

The worshippers applauded their fetich, crying, *Oricha ó, o ti kpa ó.* ("The fetich has killed him.") But their triumph was not of long duration. A few days afterward the monster floated like an inert mass on the surface of the water. The fetich priests, after a solemn funeral, thrust the body into its den under the bushes, and there it remained.

Chango.

After Obatala the most celebrated god is Chango, sirnamed *Jakuta.*

Among all his brother-gods he is the most powerful and has the most adorers. His dwelling is above the firmament, and he has a great retinue in this immense palace with its gates of bronze; he owns numbers of horses, and amuses himself with fishing and hunting. His brother Ugun, the god of war, furnishes him with chains of fire (*manamana*, lightning) which he hurls from high heaven upon his enemies. He is always accompanied by his messenger (*Ara*, thunder), who sends forth with loud noise *manamana* (the chain of fire). His three wives are beside him: *Oya* with her messenger *Aféfé* (the wind); *Oba* and *Ochun* with the bow and the sword of the god. *Biri* (darkness) entirely envelops the god and his companions.

Such is Chango of the negro theogony. The Chango that is now venerated is not the same; although he has the same attributes as the ancient Chango, he is of more recent date. According to the fetich-priests this god was a very cruel, wicked king of Yorouba. War, injustice, theft, murder, and all sorts of violence accompanied him. He despised the ancients and the fetich-priests, utterly ig-

THE GOD CHANGO, AND THE THREE GODDESSES OF YORUBA.

nored the national, civil, and religious traditions, thereby rendering himself odious to everybody. The government of the Yorouba countries is a patriarchal monarchy. The king as well as the chiefs of the province and of families have each their associates who aid and counsel them in the government. Under them the ancients are the guardians of the customs and usages of the nation; they never allow them to be in the least changed or altered, and are thus a barrier against the despotism of the kings, the governors of the province, and the heads of families. When a sovereign oversteps his rights and contemns the warnings of the ancients, the chiefs of the city police inform the king that he has served them long enough and beg him to rest from the cares of office. To this end they send him parrots' eggs in a gourd, which among the blacks has the same significance as the silk cord of the Turks. Hence comes the popular superstition that whoever looks upon the parrot's eggs becomes immediately blind. The sending of the parrot's egg signifies: Choose the kind of death which would be the easiest to you; otherwise we will choose for you.

Chango perfectly understood the message. He tried to assemble his people, but they having been warned in time, no one dared disobey the ancients. Then he chose exile. Only one of his wives and a faithful slave consented to follow him. Chango escaped in the night in order to reach Tapa, his mother's country, beyond the Niger. During his nocturnal flight his wife abandoned him. He and his faithful slave lost their way in the forest and could not regain the road. Having wandered at random for several days, Chango, worn out with fatigue, exhausted with hunger, a prey to despair, said to his companion, "Wait here; I will return, and we will continue our journey." The slave waited a long time, and when his master did not return he went to seek him, and found him dead hanging to a tree called *Ayan*. Not knowing what to do, the poor negro ran out into the open country, and soon met the people coming from the market of Oyo. He called them to aid him in rendering the last services to the deceased, and soon the news spread in Oyo that the king had hung himself,—for, according to the customs of the blacks, it is not permitted to say, "The king is dead;" you must say, *Ilé bajé, oba ti lo.* ("The earth is lost, the king has departed.") Among the negroes royalty is deified; kings are supposed to be of the race of gods, and after death become demi-gods.

When the ancients, the jealous preservers of the ancient traditions, heard that Chango had hung himself, they were terrified. The religious and political system was about to receive a mortal blow, and they would be accused of having caused the death of the king. They must at any cost extricate themselves from such terrible straits. They assembled as quickly as possible the secret police and the fetich-priests, and they hastened to the place where Chango hung himself. They buried him with a long iron chain, the end of which they left protruding above the ground; and having built a fetich-hut over the place, they prostrated themselves before it and cried out, *Oluwa wa Chango, O wo ilé lo.* ("Our master Chango has descended into the earth.") *O di Oricha.* ("He has become a fetich; he lives among the dead.")

They left the fetich-priests to worship the new god; but in the city it was said, "It is false. Chango killed himself; he hung himself." *Chango so.* The ancients and their partisans replied:

Chango ko so, o di oricha: "Chango did not hang himself, he has become oricha."

As the contending party was numerous and the people appeared incredulous, the ancients ordered the police to resort to the most severe measures to remove the evidences of Chango's self-destruction. Profiting by a storm, they set fire to the city and proclaimed, "Chango, who has become a fetich, is angry; he has not hung himself; to punish you he has hurled upon you thunderbolts."

It was announced that to honor and appease Chango sacrifices would be offered. Men and women were immolated in great numbers; terror was at its height, and all cried:

Chango ko so, ko so : "Chango has not hung himself."

This incident gave to Chango the additional name of *Obakoso*, king of the city of Ikoso, which is the place where he hung himself and was buried. It soon became a city, for the people gradually came to live near the new temple, and built houses for the fetich-priests.

The new sovereigns of Yorouba come to Ikoso on the day of their consecration to receive the sword of Chango, the insignia of their executive power.

Finally the fetich-priests of Chango adopted the ancient legends of the original Chango, and with those of their new god formed an odd mixture of ridiculous fables, of which the following is a specimen :

Chango before leaving this earth had obtained from his father, Obatala, a powerful medicine which gave him the power to overcome all his enemies. He immediately tried the effect of the medicine, and confided to the care of his wife Oya all that remained of it. She too, in secret, tasted the medicine. The next morning the king's council, terrified to see flames burst from the king's mouth at every word, fled in dismay. Oya also frightened away all her women and all the people of the palace, for she too sent forth flames from her mouth at every word. Soon afterward Chango, feeling that he had become equal to his father, took his three wives, Oya, Ochun, and Oba, and stamping his foot on the ground, which opened, he descended with them holding on to a long chain; then the earth closed up again, leaving the end of the chain above the ground. Since that time Chango has returned to earth many times and in several places; in memory of which, fetich temples and colleges have been established.

Near the mission at Porto-Novo is a place celebrated

for one of these visits of Chango. For a great many years a fetich-temple and a college stood there. The temple has been deserted for some time, one fetich-priest alone taking care of it. The college, inconvenienced by being so near us, has been removed elsewhere. The following is the legend which commemorates this fact. Chango, who had been brutal and wicked during his life, became still more violent and brutal after death. Very often every year he flew into a violent rage, hurling thunderbolts and only calming himself to return again to his fury.

One day *Oya* (Niger), his cherished wife, dreading his fury, fled to her brother *Olokun* (the sea). Chango, hearing of it, was furious and resolved to revenge himself. He followed the sun in its course to where the sea and the sky meet. Here, say the blacks, is where the white people come in their ships and find all the things with which they fill them. Chango descended into the empire of his brother Olokun. Oya, terrified, ran to her sister Olosa (a lagoon which connects with the Niger and with the ocean), pursued by Chango making a frightful fracas. This is why there are no trees on the border of the lagoon; they were uprooted at this time and cast away.

Oya fled again from her sister's abode and took refuge with Houésé, an inhabitant of the country,—who has since become a fetich, thanks to the supernatural power communicated to him by Oya by means of the medicines which she had stolen from her husband at the time of her flight. Houésé warmly took up the defence of Oya, and a memorable struggle ensued. The combat began near the water where Houésé's hut was built. Chango seized his canoe as a club; Houésé, feeling a divine force, armed himself with a tree: canoe and tree flew into splinters. During the fight Oya fled to *Lokoro*.[*] Then the two ad-

---

[*] In the neighborhood of Porto Novo.

versaries closed with each other, and with flames bursting from their mouths fought with terrible fury; their feet dug deep furrows in the earth, like those still seen near the missions, which are the deep excavations made by the rains. Finally, neither adversary being able to conquer, Chango, exhausted with fatigue, disgraced and enraged, stamped on the ground, redescended among the dead, and from there entered into higher Olympus. Oya (the Niger) remained at Lokoro, where a temple was built to her. The goddess is here represented by a statue with nine heads, one of which is surrounded by the other eight—a symbol, no doubt, of the Niger and its tributaries. Houésé has his temple on the place where he descended alive into the earth.

The fetich-priests of Chango could not meet those of Houésé without giving battle in memory of the memorable struggle of their respective chiefs, each party maintaining that his god is the stronger. The king of Porto-Novo, to avoid all difficulties, forbade the fetich-priests of Chango to come beyond the boundary-lines, and commanded those of Houésé to remain on the other side at the time of the fetich-feasts. The fetich-priests of Chango wear a sack, the emblem of pillage, in honor of the pillaging virtues of their master. At certain times they are privileged to steal the hens and goats they meet in the street. When a house is struck by lightning they have a right to pillage it, and to impose a fine on the victims of the fire, in order to complete the vengeance of Chango, who it is supposed strikes only the guilty with lightning. And to prove the guilt of the household they search for thunderbolts,—and always find them without any difficulty, as they bring them with them. At Why-dah, the mission-house was struck by lightning and reduced to ashes; the fetich-priests of Chango exacted a large fine; Father Borghero, then Superior, was put in prison, and was only released when the French merchants kindly paid the fine for the mission.

Red and white are Chango's colors. Hens and other animals are offered to him, and even human victims.

Oya, Ochun, Oba.

After Chango were born Oya, Ochun, and Oba, all three wives of Chango. Oya (the Niger) had for slave *Aféfé* (the wind). When her husband thundered, she preceded him with her messenger *Aféfé*. *Ochun* and *Oba* (two rivers of Yorouba) followed their spouse: one carried his bow, the other his sword. Finally *Biri* (darkness), Chango's slave, accompanied his master, whose will and vengeance *Ara* (the dust) executed.

Dada.

After these three goddesses was born Dada (Nature and vegetables). The symbol of the god of nature is a gourd ornamented with white porcelain shells and a ball of vegetable indigo.

Ochosi.

Ochosi, a hunter, whose symbol is a bow.

Ajé Chaluga.

Ajé Chaluga, the African Mercury, god of riches, son of the Ocean, has for emblem a large sea-shell, before which offerings are made to obtain from the god riches, which on the Slave Coast are acquired with porcelain shells, small sea-shells which take the place of gold and silver. All colors are consecrated to him.

Ugun.

Ugun (a river of Yorouba which flows into the sea near Lagos) is the god of war and the chase. Any piece of iron may serve for his symbol. I have heard the blacks swear, saying, "May Ugun kill me if I lie!" and they kiss the iron of their sabre. Warriors, hunters, blacksmiths, and all those who use iron tools offer sacrifices to him. He is especially the god of blacksmiths, for Ugun is the negro Vulcan and furnishes Chango with the thunderbolts, *manamana*, and the red-hot iron chains which this

A HUMAN SACRIFICE TO UGUN, THE GOD OF WAR (SLAVE COAST).

Oké.

Oké is the god of the mountains. His symbol is a stone.

Oricha-Oko.

Oricha-Oko, the god of the fields and of agriculture, has for emblem a long iron bar. This god, the brother and friend of Chango, is much honored among the blacks and has a great many temples and fetich priests and priestesses. The bees are his messengers.

Champana.

Champana (the small-pox) is a diseased, deformed god. The gods being one day assembled at the palace of Obatala, the father of the gods, they were invited to dance. Champana stumbled and fell, thus exposing himself to the raillery of his sister-goddesses. In his anger and disgust he tried to communicate to them the small-pox, but Obatala repulsed him with his lance and drove him away. Since that time he inhabits the deserts and forests; his temples are always built outside of the city, in a wood or grove. Mosquitoes and flies are his messengers. His symbol is a large stick marked with red and white spots. He is the most dreaded of all the fetiches.

Orun and Ochu.

Orun and Ochu (the sun and the moon) are obsolete gods to whom sacrifices are no longer offered. The sun and the moon had innumerable children. The young suns wished to follow their father, but he, jealous of his power and wishing to be alone, fell upon his children and tried to kill them. Then the children precipitated themselves on the earth and took refuge with Iyemoja, their grandmother, who changed them into fish and took them to her bosom. Some of them remained with her; others went in great numbers to Olosa (the lagoon) and to Olokun (the sea). Only the daughters of the moon found grace with their father; they accompany their mother at night.

But it sometimes happens that the sun leaves his course to follow the moon and maltreat her. Then the blacks come out and shriek and beat drums to frighten the sun and oblige him to leave the moon in peace. This uproar takes place every time there is an eclipse of the moon. Cries of "Away! Begone! Leave her!" fill the air until peace is re-established in the heavens.

3. *Ifa.*

Having enumerated the inferior divinities dependent upon the two first superior divinities, we come to the third. After Obatala and Odudua, Ifa is the fetich the most honored among the blacks. He is the revealer of future events, the patron of marriage and of birth. He is also called Bango (the god of the palm-nut), because sixteen palm-nuts are used to consult the god and obtain an answer. The city sacred to Ifa is Ado, built on an immense rock.

Nothing is done without consulting him, and they always act in accordance with his answer. He is the messenger and interpreter of the gods; it is through his ministry that the fetiches manifest their will, and that man makes known his wants.

The legend represents that Ifa came from the city of Ifé, but does not clearly indicate who were his father and mother; it seems that he was, like the others, the son of Obatala and Odudua. He is the benefactor of humanity, the god of wisdom. He left the city of Ifa, which would not listen to him, and roamed the earth to instruct man in the arts, and above all in the knowledge of the future. Several instances are related which show that if he is the god of wisdom, he is not the god of chastity. However, immorality is the distinctive characteristic of the pagan gods.

After many peregrinations and adventures, Ifa finally

established himself at Ado and planted on the rock a palm-nut which produced sixteen palm-trees from the same root.

The legend relates that when Olokun, the god of the sea, destroyed nearly all mankind by the floods, there remained only a few that Obatala had saved by drawing them up to heaven with a long chain. Then Ifa and Odudua descended to earth to render it habitable again; this is why Ifa and Odudua are held in such great veneration. But on this point tradition becomes vague and sometimes contradictory.

Another tradition, which seems the same as the above under a different form, relates that the first people sent to establish themselves at Yorouba (it is not said from whence or by whom) were obliged to walk for a long time in the water which covered the earth. He by whom they were sent had given them a little earth and a palm-nut in a piece of linen, and also a hen. When they had travelled a long time, as the earth was still covered with water they threw the palm-nut into the waves, which immediately produced a palm-tree with sixteen roots. The blacks climbed into the tree to rest. Then they threw the earth into the waves, and it immediately formed a small hill, upon which the hen flew and began to scratch and spread the earth with her claws. Gradually the earth expanded, the water disappeared, the travellers were able to continue their journey, and they arrived at Yorouba, where they established themselves. We see in this legend the deified earth and the sacred tree: the palm recalls Ifa.

Ifa taught men the art of consulting fate to know the future and the will of the gods in the following manner: In the beginning, when there were very few men in the universe, Ifa and the other gods had not as now an abundance of presents and sacrifices. The gods were obliged to contrive means of satisfying their desires. Ifa among others devoted himself to fishing. One day, exhausted

with fatigue, he addressed himself to Elegba, the most
acute and crafty of the genii, as well as the wickedest, and who like himself suffered want in wandering
through the deserts with his companions the spirits.
Elegba, consulted as to the best means of ameliorating
their mutual conditions, replied that if he had sixteen
dates from the two palm-trees that Olorun Olodoumaré
(the all-powerful god) had entrusted to the care of man,
he could teach him the art of knowing the future and of
propitiating the gods, in order that he might have part in
the sacrifices which were offered them. But before confiding to him his secret, Elegba stipulated that he should
have the first choice of the sacrifices offered to the gods.
Ifa accepted these conditions and promised to make
Elegba's wishes respected; and this custom is still observed. Ifa went in search of Orungan, the chief of men,
and made him understand the advantage it would be to
him to know the future and the will of the gods, and thus
draw upon himself their favor and avoid their anger.
The chief allowed himself to be persuaded; he and his
wife Arichabii ran to gather the sixteen dates necessary
to the magical operations, but they could not reach them,
the trees were so high. God had commanded them to be
very careful of these two trees, above all never to allow
the monkeys to climb them or injure them. They withdrew a short distance, and permitted the monkeys to come
and throw down the required sixteen dates. These animals, having already been tempted for a long time by the
sight of the sixteen full ripe dates, leaped with one bound
into the trees and began to eat the red pulp which surrounds the stone of the date, and threw upon the ground
the remains of the despoiled fruit. Orungan and his wife
gathered them up; the woman tied them up in a piece of
cloth and fastened the package on her back in the way
the negresses carry their chidlren. Then they both
tried to chase away the monkeys, but could not, for
when driven from one tree they jumped into the other,

breaking and injuring at pleasure the two beautiful palm-trees.

Ifa taught Orungan how to make use of these dates in consulting fate, and he selected one of his faithful Ochougbolu and explained to him the method and ceremonial to be observed in consulting the future. In memory of this tradition, when they wish to consult fate or to make a grand ceremonious feast in honor of Ifa in the grove sacred to this god, the mother or wife of him for whom the god is consulted carries in a cloth on her back the sixteen sacred nuts, and the fetich-priest before commencing the ceremony salutes Orungan and his wife, saying, *Orungan ajuba ô!* ("Orungan, I salute you.") *Orichabii ajuba ô!* ("Orichabii, I salute you.")

Then he offers sacrifice to Ifa, of which the dates are the symbol. Finally he places before the god a small board on which are marked sixteen figures, each having a certain number of points. These figures are very similar to playing-cards used by fortune-tellers. The fetich priests use them in almost the same way, bringing out at will good or bad fortune according as they deem it expedient to better dupe the fool who comes to consult them. When he has found the desired figure, he begins to explain whether the enterprise in question will succeed or not, the sacrifices to be offered, the things to be avoided. It is well understood that the higher the price paid the greater the inspiration of the fetich-priest, for there are large and small games.

Ifa is the most venerated of all the gods; his oracle is the most consulted, and his numerous priests form the first sacerdotal order. They are always dressed in white, and shave the head and the body. They offer sacrifices and libations to Ifa, and on certain important occasions **they** immolate to him human victims.

II. Demi-Gods.

Deification of Humanity.

Besides the principal divinities of which we have just spoken, there are many others less important which cannot be enumerated, for their number daily increases. A family establishes itself near a river, a forest, a rock, or mountain; imagination aided by the fetich-priests soon creates a belief in a demi-god, a tutelary genius of the place, and thus a new divinity makes its appearance in the negro pantheon, and it is not long before it has its legend also.

The worship of the dead has greatly aided in augmenting the number of the gods. Joined to the worship of nature is that of humanity. The descendants from generation to generation offer presents and sacrifices on the tomb of their ancestor, and end by adoring him as a local divinity, the origin of which becomes more and more obscure and consequently more and more venerable. This occurred at Porto-Novo in the case of the chiefs of families in various parts of the city, of whom the inhabitants are the real descendants.

The blacks also render divine honors to those men whom they believe to have been raised after their death to a high degree of power which renders them equal to the gods. This honor is given not to those who were celebrated for their virtues and their good deeds, but to rascals who have rendered themselves odious by exacting enormous fines and are stained with all sorts of infamous crimes. These are for the most part the pillaging kings and princes who have ravaged and devastated countries, destroyed entire cities, who were the terror of their subjects and of their own families.

Fetichism.

Such is Ajahuto, whose temple is in the palace of the king of Porto-Novo; he was one of their former princes, who killed his father-in-law. When a young girl is dedicated to his worship she is obliged to remain always a virgin; she has precedence of all the chiefs of Porto-Novo; she alone does not prostrate herself before the king. She takes care of the temple of the demi-gods, to whom she offers the sacrifices. The last one was executed for having failed in her duty, and no one has yet been found to replace her. Every year human victims are immolated to Ajahuto.

Another demi-god famous at Dahomey is Adanlosan, another celebrated king. The blacks maintain that he is not dead, but that he became a fetich in life, and that he often comes to dwell in his palace at Abomey. The reigning king is very careful to do nothing without consulting him; but both he and the fetich-priests know how much of this to believe.

Adanlosan was a very cruel chief, and at the same time a dreaded warrior. He it was who closed with baskets filled with sand the estuary through which the waters of the great lake Nokumé flowed to the sea at Kotonou, and forced them into the lagoon which empties near Lagos. On the canal thus constructed he passed with his army to destroy the city of Tocpo, not far from Badagry. The mission has a beautiful farm on the ruins of this city. This king finally made himself so odious to his family and his subjects that the ancients resolved to rid themselves of him. An opportunity of carrying out their resolution presented itself in the following manner: One day his brother's little son while amusing himself threw a stone which struck Adanlosan. He immediately ordered the child to be beheaded. His father as well as all those present begged and implored mercy for the boy, but the king was inflexible and commanded his orders to be executed. Then the child's father fell upon the chief, threw him from his throne, and a terrible struggle ensued;

but no one came to the aid of the detested tyrant; they bound him hand and foot and shut him in a small room in the palace, the entrance to which they walled up and left him to die of hunger. The report was soon spread abroad that the king being old had become a fetich and shut himself up in this room in order to remain among his people and protect them.

The worship and homage rendered to the dead is exactly the same as that given to the fetiches. They have their temples and their priests; sacrifices are offered to them, sometimes human sacrifices.

III. Genii.

After the gods and the demi-gods come the spirits or genii. The genii are very numerous; some are good and some bad spirits. A certain number serve as messengers to the gods and demi-gods; some are considered nearly as powerful as the gods themselves and have authority over lesser spirits who are their messengers, and these in turn command others, forming a hierarchy which is not very defined. The more ordinary spirits dwell in the forests and deserts.

GOOD GENII.

The good or protecting genii are those who are considered as being well disposed towards man. They are deputed by Obatala, the father of the gods, to take care of the different parts of the universe. Although good, they are subject to anger; their moods are very changeable, and they are exacting in their service.

Aroni.

Aroni is the genius of the forest, and is said also to be skilful in medicine. He is not very good, and is very capricious and much dreaded by those who do not know his character. This genius appears under the human form with a dog's head and only one foot. At other times he manifests himself by a whirlwind which rushes through the forest carrying the leaves before it. Whoever meets him in the forest and has the misfortune to run is devoured. But to him who stands his ground and looks fearlessly at him the monster becomes as gentle as a lamb. He conducts the happy mortal to his palace in the depths of the woods, where for several months he takes the greatest care of his guest, teaches him all sorts of remedies, explains to him the properties of the various barks and roots, and finally confers upon him the title of doctor, giving him for a diploma a hair from his tail.

I met one day an old fetich-priest who called himself a disciple of Aroni; he showed me his diploma, a hair from the back of a wild-boar. He said he knew everything about medicine, offered to secure me against fever forever, assuring me that for the future I would be as strong as iron-wood, and that I would live till the end of time like an old moss-covered trunk of a tree. The rogue did not ask much for his services: only two bags of shells to buy his food during the two weeks which he would have to spend in the forest in order to procure all the roots and bark necessary to change me into wood; then a bag of shells to purchase a perfectly new black pot, a sheep to consecrate the pot by a solemn sacrifice, and finally of course a bottle of rum to give him strength to dance around the pot during the mysterious operations. I dismissed the rascal with his promises, his medicine, and his diploma, advising him to concoct a remedy to rejuvenate his wrinkled skin, and he went off grinning.

Eléda.

Every man has three genii, or protecting spirits. The first is Eléda, who dwells in the head, which he guides. A hen is offered in sacrifice to him; a little of the blood of the fowl, mixed with oil is rubbed on the forehead. This is the genius' portion; then the hen is eaten. A small package of white shells is his symbol.

Ojehun or Opin ijéhun.

This second genius has his habitation in the region of the stomach. He is the most favored, and the one to which the blacks pay the most attention because of his position. Opin ijéhun, whose name signifies "he who takes part in the food," is also the genius who keeps up the

fire; he never permits this precious element which serves in preparing the food to be extinguished, naturally finding this to his advantage.

His messenger is Ebi (hunger). When the negro is idle and lazy, Ebi pinches his stomach, and the lazy negro is obliged to go to work to gain wherewith to satisfy Ebi and his master. The negro often when begging will point to his empty stomach, saying, *Ebi pa mi* ("Ebi kills me"). Of course no particular sacrifices are offered to Ojehun; he is served every day as well as the means of the negro will permit.

Ipori.

Ipori, the third protecting genius, takes up his abode in the great toe. This genius is very poorly provided for. Sacrifices are very rarely offered to him, only when the negro is about to undertake an important journey. Then he does homage to his toe by rubbing on it a little chicken-blood and oil, and the genius is satisfied. The negro then sets out with his three genii, and he cannot fail to bring back intact his head, feet, and stomach.

Alayoré.

Alayoré is the protector of the hearth; he has the care of the house. Armed with a stick or sword, according to the fancy of the master, it is his office to drive away from the house evil spirits, particularly the terrible Elegba, who stands at the door under the wretched thatched roof.

Osanyin.

Osanyin, the genius of medicine, is of all the genii the most highly esteemed and the most frequently consulted. His symbol, an iron rod surmounted by a figure of a bird, is found in the court-yard of all the houses, generally at the foot of a tree. When the fetich-priest consults him, the genius answers in a voice very like a little chicken or bird. Of course it is the fetich-priest who asks the question and gives the answer. The crafty charlatan never fails to do it very adroitly.

One day while travelling I went into a fetich-cabin to rest; there were three other little cabins near mine in the same sacred grove, not far from a village called Ipobita. Scarcely had I entered when a band of negroes and a fetich-priest arrived; they came to consult the god of the genius of medicine about a poor man who had hydrophobia. The sick man remained with the others outside of the grove; the fetich-priest entered one of the sacred cabins near the one I occupied. It seemed that the genius was out walking in the neighboring forest, for the fetich-priest began calling him by ringing a little iron bell. After a few strokes of the bell a low whispering was heard in the distance. It gradually drew near until it was heard in the cabin where the fetich-priest was. (This whispering sound is made by putting a leaf or a blade of grass between the teeth and the lower lip.) Then a dialogue began, the genius whispering, the fetich-priest answering; finally he explained to the faithful, lying outside with their faces to the ground, what the god said. At first the medicine-genius asked a very high price for curing the sick man. He replied that he could not pay; that he was poor; that he had already spent much, and that he could no longer work. After a great deal of parley everything was arranged, and the fetich-priest left the temple satisfied. They drank freely of palm-wine, a full gourd of which was offered to the genius and poured in front of his symbol.

Aïdowedo.

Aïdowedo (the rainbow) is a genius held in great veneration at Porto-Novo. In Yorouba he is called Ochumaré. The temples dedicated to this genius are painted in all the colors of the rainbow, and in the middle of the prism a serpent is drawn. This genius is a large serpent; he only appears when he wants to drink, and then he

TEMPLE OF THE FETICH-SERPENTS.

rests his tail on the ground and thrusts his mouth into the water. He who finds the excrement of this serpent is rich forever, for with this talisman he can change grains of corn into shells, which pass for money.

The blacks are firmly persuaded of this. One day I tried to disabuse my little negroes of this idea, and explained to them by means of a prism the way the rainbow was formed. A negro man who was present, seeing the colors reflected in the prism and paying no attention to the explanation, believed that with this piece of glass I could bring down the rainbow at will. He became firmly convinced of this, and told that he now knew how without being in trade we always had shells to buy food and build houses; for, said he, he has shown me how to bring Ochoumaré into my room. Soon several negroes came to beg of me some of the precious excrement. I had the greatest difficulty in getting rid of them, and they went away persuaded that I wished to keep the much-desired substance all to myself.

Among the Yoroubas the boa-constrictor, called Eré, is supposed to be the messenger of this serpent-genius.

When a boa is declared to be the messenger of the god, it is not permitted to kill him; on the contrary, presents must be made to him. Leaves from the fetich palm-tree of Ifa scattered about indicate to the devotees the place chosen by the monster for his dwelling. Woe to the cabins that are in his neighborhood! for hens, goats, sheep, and even little children are in danger.

One of these subordinate divinities, driven by fire from the bushes where he lived, took up his new abode in a thicket near the house of one of our Christians. The neighboring fetich-priest declared that the serpent was sacred, and scattered palm-leaves in front of the thicket, thus proclaiming the thicket sacred also. The neighboring blacks offered hens in sacrifice to the new god; but as these sacrifices were not frequent, the god came out at night and devoured all animals that had not been carefully locked up. All the hens or goats that approached too near the thicket during the day forfeited their lives by their temerity. My Christian dared not rid himself of his neighbor, having a certain superstitious fear of him. I advised him to offer the monster a sacrifice difficult to digest : and by this means he was disposed of.

At Dahomey and at Porto-Novo a small poisonless, very inoffensive species of boa called *Dangbé* (*dan*, serpent, *gbé*, life) is consecrated to this genius, and is considered his messenger. This serpent has his temples and fetich-priests ; it is forbidden to kill him under the most severe penalties ; and but for the pigs, an unsuperstitious race which devour the serpents in great numbers, it would be impossible to keep any domestic animals.

Dangbé in his turn has white ants for messengers, doubtless for his lesser wants. A hillock of white ants is often seen surrounded with palm-leaves to indicate that the inhabitants are now in the service of Dangbé. If a negro sees a serpent coming out of an ant-hill, he runs at once and reports this to the fetich-priest, who immediately brings palm-leaves, which he scatters about the mound.

EVIL GENII.

Elegba or Echu.

The chief of all the evil genii, the wickedest as well as the most dreaded, is *Echu*, a word signifying " the rejected." He is also called *Elegba* or *Elegbara*, " the strong," and again *Ongogo Ogo*, "the genius of the knotted stick."

To protect themselves against his wickedness, the blacks keep in their houses the idol *Olaroza*, the protecting genius of the house, who, armed with a stick or sword, guards the entrance. But in order to ward off his cruelty, when obliged to go out to attend to business, they never fail to give him his share in all the sacrifices. When a negro wishes to revenge himself on an enemy, he makes a generous offering to Elegba, presenting him with a copi-

ous draught of tafia—palm-wine. Elegba then becomes enraged, and if the enemy is not well protected with amulets he is in great danger.

This is the evil genius who, by himself or with his companions, urges men to sin, and above all excites in them shameful passions. Often when negroes are punished for theft or other misdeeds they excuse themselves, saying, *Echou l'o ti mi.* ("Echou made me do it.")

The image of this wicked genius is placed in front of the houses, in all the parks, and on all the roads.

Elegba is represented seated, with his hands on his knees, and perfectly nude, under a sort of roof made of palm-leaves. The idol is made of clay in human form, with an enormous head; birds' feathers form the hair, two shells represent the eyes, and shells also form the teeth, giving him a horrible appearance.

On grand occasions he is saturated with hen's blood and palm-oil, which gives him a still more hideous and disgusting appearance. To complete the worthy decoration of this ignoble symbol of the African Priapus, the handle of an old pickaxe or a large knotted stick is placed in front of him. The vultures, his messengers, fortunately eat the hens, dogs, and other victims that are immolated to him, which would otherwise poison the air.

His principal temple is at Woro near Badagry, in the midst of a charming grove of palms and other beautiful trees. Near the lagoon, where a grand fair is held, the ground is strewn with shells which the blacks scatter there as offerings to Elegba, in order that he may not interfere with them. Once a year the fetich-priests of Elegba gather the shells to buy a slave to sacrifice to him, and brandy to inspirit the dancers; what remains is for the fetich-priest.

The following instance shows Elegba's inclination to make mischief: Jealous of the perfect harmony which existed between two neighbors, he determined to set them at variance. For this purpose he put on a cap one side of which was pure white, the other bright red, and going out he passed between the two friends, who were working in their fields, saluted them and passed on.

When he had passed one said to the other:

"What a beautiful white cap!"

"Not at all," replied the other; "it is a lovely red one."

From this the dispute became so animated between the two men that one broke the head of the other with a pickaxe.

Chougoudou.

When the blacks wish to guard a place and inspire a fear of it, so that no one will dare approach it at night, the fetich-priests dig there a small hole in which they immolate to the evil spirits a hen or some animal, often even a human victim in order to have a stronger and more wicked spirit. They then cover the victim with earth, forming a sort of round tomb, on the top of which they place a vessel for the pittance offered to the spirits, who, thus duly installed, guard the post. The blacks have a great fear of the Chougoudou, and will never pass at night the places where they are, for fear of being maltreated by the spirits. The palace of the king of Porto-Novo is under the powerful protection of a Chougoudou.

Genii of the Trees.

The iroko and several other kinds of trees are considered to be the dwelling-place of evil spirits. The tree supposed to be haunted by a spirit is marked by a circle of palm-leaves; a path leads to it, and earthen vessels and human bones are laid against its trunk. Thanks to this superstition, the large beautiful trees which adorn the African cities and villages are spared. However, if they want to cut one down which is not in the city, they can with sacrifices of hens and oil and with the help of money make the genius leave the tree and install him elsewhere. If a negro goes to the forest to cut down one of the trees

FETICH TREE.

usually chosen by the genii, through fear of the evil spirit he does homage to his good genius by rubbing a little oil on his forehead, and then fearlessly cuts down the tree. At Porto-Novo a European, not knowing the customs of the country, had a large tree cut down which interfered with his building. The negro who did the work, believing that the white man had authority for it and that the ceremony of transfer had taken place, cut down the tree without troubling himself about it. But the king, hearing of it, had the negro beheaded without trial, saying, if the white man did not know the customs of the country, the negro ought to know them.

The blacks believe that the sorcerers, called Ajé, assemble at night at the foot of these trees to do homage to the spirits living in them. When these sorcerers wish to revenge themselves, the spirit places at their disposition his messenger the owl, who, directed by an inferior spirit, goes to the house of the person the sorcerer wishes to kill, and eats his heart out in the night. When this bird is seen in the house, it is for the purpose of killing some one. If he can be caught and his legs and wings broken, it is believed that the same harm is thus inflicted on the sorcerer who sent him.

This is one of the most deep-rooted superstitions in the minds of the blacks, and is the cause of much vengeance and many crimes; even the Christians find great difficulty in ridding themselves of it.

The old women are often accused of being ajé. Many of the poor old creatures are submitted to the test of Oncé, condemned to death and executed the following night. The most curious part of it is that they often believe they have really committed the crime of which they are accused. Doubtless to be revenged or to gain a sum of money they have gone to the foot of the sacred tree and asked the genius to send his messenger to kill such or such a one, and when the victim dies they believe that the owl has gradually eaten his heart.

Among the blacks, white magic, the object of which is to do good, ward off evil, and cure the sick, is permitted; but black or wicked magic is forbidden under pain of death. Everybody accused of black magic and found guilty according to the tests used in the country is executed, and the executioners eat his heart in retaliation. Often the blacks do not wait for legal proof to revenge themselves. Recently in a little village near Lagos an old woman was murdered with unheard-of cruelty by a negro and negress who accused her of being ajé, and of having made the owl eat the hearts of their children. All the blacks were convinced that the old woman was a sorcerer, and this belief assured them they were justified in murdering her.

Abiku.

There is another genius, called Abiku, who, instead of perching on the trees, takes up his abode in the human body. Children who die between ten and eleven years of age are also called Abiku, and are never buried, but are thrown in the bushes. It is supposed that the child and the spirit are thus punished.

There are great numbers of evil spirits called Abiku and Eléré, who inhabit the forests and deserts, suffering want and having a great desire to share the good things enjoyed by mortals in this world. With this object they watch for the entrance of the soul into the body formed by Obatala. One of them at once installs himself therein, promising the other spirits, his companions, to share with them the good things which he is about to enjoy in this world.

When a child cries and suffers more than usual, the blacks believe that the spirits, companions of the one who is in the child, are hurting him in order to get more food. If the child grows thin and puny, it is because the evil spirits steal all the nourishment he takes. In order to cheat these spirits a sacrifice is offered to them, and while they are feasting on the offering, little bells are placed on the feet of the child, the tinkling of which drives off the evil spirits and keeps them at a distance. It is not unusual to see a little negro with his ankles loaded with little bells and pieces of old iron, an insupportable burden to the poor little child.

If the child supposed to be haunted by an evil spirit becomes dangerously ill, the mother makes an incision in the body and puts spices in it, believing she will thus make the spirit suffer and force him to leave the child.

If the child dies, the body is thrown on the dirt-heap to be devoured by wild beasts. The mother will often madly mutilate the corpse of the poor child, pound it with stones, cut off an arm or an ear, and threaten to beat the evil spirit, calling him wretch, thief, etc.

Belief in this error is still more strengthened by the fact that sometimes another child is born with the marks of the wounds made on the body of his elder brother, because the image of them remains engraved in the memory of the mother; but the blacks will not accept this explanation, and hold to their superstitions, which the fetich-priests have every interest in fostering.

These evil spirits have great power over the bodies they possess. It is related, in illustration of this, that a woman was in the habit of leaving her infant on a mat in the cabin while she went to market, and although the door was locked all the food she left there disappeared. Moreover a neighboring peddler reclaimed from the mother the shells which she said her son had come to borrow from her. The negress showed her that the child sleeping on the mat was too young to walk. Nevertheless the neighbor affirmed that she had seen him, a half-grown child, leave the house, come to her, take the shells, buy food, and then return to the house. To search into this mystery the father carefully concealed himself in the cabin. When the woman had gone out as usual, firmly fastening the door, the little child stood up. became suddenly a big boy, searched about, found the shells, and was just going out, when the father appeared. At sight of him the rogue became again a little child, crying and sobbing. Such are the ridiculous stories which circulate among the blacks and retain them in their superstitions.

Ibeji.

When a woman has twin children, they are not killed at Porto-Novo, as is the practice at Benin, but the blacks think that these children have for companions genii similar to those which animate a small species of monkey common in the forests of Guinea. When these children are grown they cannot eat the flesh of monkeys, and meanwhile the mother makes offerings of bananas and other dainties to the monkeys to propitiate them.

If one of the twins is ill, the mother consults the fetich-priest, who invariably commands her to make the customary sacrifice to the spirits, that they may leave the child in peace. The negress, with her basket well filled with wine, nuts, bananas, and other dainties loved by the spirits, goes with her companions to the fetich-priests to make her offering. It is laid at the foot of a tree; the fetich-priest evokes the spirits, and when they manifest their presence, all retire to allow them to eat in peace. Presently they return to see if the genii have found the offering to their taste. When everything has disappeared, it presages favorably for the health of the child. It is needless to say that the spirit who accepts the sacrifice is a spirit of flesh and blood, who, notified beforehand, conceals himself in a convenient place near by.

The African Wind, and Genius of the Locusts.

The genius of the African wind, called Oyé, dwells with the genius of the locusts in the grand official temple of Elegba, chief of the evil genii. This palace is built on Mount Igbeti, near the banks of the Niger. Every year the grand fetich-priests of Elegba open the great bronze doors of the temple, and offer a solemn sacrifice to all the genii and to their chief. Then the African wind rises and covers the earth; the locusts take flight wherever the spirits impel them, then at command of the genius Oyé the African wind and the locusts return to the temple.

Genius of Togo.

There is near Porto-Novo a lagoon called Togo, which has served as a legal test from the time of the ancient kings. It has lost much of its prestige in the present day, being replaced by Oncé.

The test consists in conducting the accused to the middle of the lagoon, to a spot known to the fetich-priest charged with this service. The accused is thrown into the water. If he float, he is taken into the canoe and declared innocent; if he sink, the genius Togo is said to have killed him, and the next day his body is found near the bank, on a bamboo raft, where the god placed it.

The following legend gives authority for this belief. A poor negress went to gather wood on the banks of the lagoon, to buy food for herself and her two children; but notwithstanding all her efforts she with great difficulty maintained her unfortunate existence. She denied herself in all things for her pretty little negroes, who, ignorant of their mother's miseries, amused themselves on the banks of the lagoon. One day the negress missed her dear children. She went about inconsolable, seeking them everywhere, making the air resound with their names, but in vain. Time assuaged not her sorrow, and she came every day to weep on the banks of the lagoon. The genius of Togo was touched by her grief. One day, to her great astonishment, she saw her two children coming toward her, their bodies half above the water, and swimming like fish.

"Weep no more, mother," they said, "for we are very happy here. The god of the lagoon had pity on thee, seeing thy difficulty in providing for us, and took us to his home, where we have fish and food in abundance. We are in a beautiful cottage under the waters. The fishes play around us, and every day is a feast. Go tell the king that the god of Togo wishes a temple to be built to him on the banks of the lagoon which he guards, and wishes that sacrifices be offered to him. The god in return will make known to him the guilt or innocence of the accused in doubtful cases. For the accused who shall be thrown into the lagoon, if he be innocent, shall not drown; but if guilty, he shall be dragged to the bottom of the lagoon, and his body cast upon the banks."

Formerly the lagoons opened every year, and all went to offer presents and sacrifices to the gods and goddesses who came to amuse themselves in the magnificent dwelling of the god of the waters. From the lagoons they passed into the sea. Having danced and amused themselves to their heart's content, the men returned home, and the sea and the lagoons closed up again. But since untruth has come upon this earth, the sea and the lagoons no longer open; the people must be content with offering sacrifice to the gods in the temples built to their honor on the banks of the waters.

Another legend accuses a wicked king of Dahomey of having put a stop to the communications of mortals with immortals. All the gods and goddesses were assembled, dancing and amusing themselves, when the king ordered his amazons to seize the goddesses, carry them off to Abomey and enlist them in their regiment; but as the amazons were about to seize the goddesses, they all disappeared, the divine dancers having changed themselves into drops of dew. Since that time this world has been deprived of seeing the immortals.

Deified Power.

Among the blacks of the Slave Coast power is deified. Kings are supposed to be descended from the demi-gods, and consecrated to them, and initiated in the secrets of the negro sanctuary. White is the official color of the first order of priests, and they wear white vestments.

To facilitate the administration of justice, the blacks have from time immemorial made religion intervene under different names. They deify the executive power, and regard the judges and the executors of the law as supernatural beings and descendants of the gods.

Judicial power is deified in Yorouba under the name of *Egungun* (bones, or the dead). Egungun appears in the streets in the form of a demon masked and fantastically dressed, walking ridiculously, and uttering loud discordant sounds; from time to time he changes the mask, sometimes taking the face of a dog, at another that of a monkey. Egungun is supposed to come from the other world to see what takes place in this, and to carry away with him all those who are disposed to trouble the living.

When there is a death in the house, Egungun and his companions, equipped like himself, never fail to come and pay their respects to the relatives and bring them news of the deceased, assuring them that he is well, and that he has gone safely through the terrible passage and arrived without accident in the country of the dead. The charitable Egungun and his satellites are loaded with presents and invited to rest; food, and above all a generous supply of rum or palm-wine, is placed in a room to which they retire, for it is death to look upon the dead while they are eating. Egungun and his companions, like the other black spirits, have good appetites; when they have eaten to their satisfaction they depart, returning thanks in deep groans to the relatives of the dead, who charge them with messages for the dear deceased, and they indicate by renewed groans that their commission shall be fulfilled.

A criminal after he is condemned to death is given over to Egungun. He cuts off the head of the victim, with which he promenades through the streets of the city. The body is thrown into the bushes, and cannot receive burial unless it is bought by the relatives. No one has a right to raise his hand against Egungun, not even the king. As to the women, they are forbidden under pain of death to say what they think of him.

Oro.

Among the Egbas executive power is deified under the name of *Oro* (tempest). Under pain of death all women young and old are obliged to believe that Oro is a powerful spirit who dwells in the firmament in company with many other genii. It goes without saying that only an exterior faith is required; that is to say, the women must be silent about what they think. As to the men and boys, they know how much of it to believe.

When a criminal is condemned to death he is handed over to Oro, who swallows him. The next day the clothes of the unfortunate wretch are seen floating on the top of a high tree, having been left there by Oro when reascending through the air after having cut off all the branches of the tree. Sometimes the formidable voices of Oro and his companions are suddenly heard resounding through the city; this is to announce that the god is abroad, and then all the women under pain of death must shut themselves in their houses. Thus affairs may be arranged without the interminable chatter of the negresses. The terrible voice of Oro is a noise made by rapidly twirling a tongue of wood attached to a string.

Oro on his feast-day appears in the shape of a monster in human form, with the face and lips besmeared with blood. His bellowing is heard in all the cities, and the blacks have great feasting in the grove of the god.

Zangbeto.

At Porto-Novo the sacred police go out only at night. They are called *Zangbeto* (the people of the night, who come from the other side of the sea). The functions of the police are the same as Oro and Egungun, only that they make a much more frightful racket; they imitate the noise of every kind of animal, with an accompaniment of an orchestra composed of all sorts of old iron which produces an indescribably infernal din. The spirit in coming from the other side of the sea assumes a ridiculous gait, and disguises himself in a large straw cone which covers him from head to foot. No one dares to venture out under pain of being severely flogged. At the time of the "Customs," when they immolate human victims, the rash mortal, who ventures out exposes himself to be sold as a slave or offered in sacrifice.

Ogboni.

There exists among the blacks a secret society, the members of which are very numerous in Yorouba and are called Ogboni. This sort of freemasonry seems to have for object the preservation of the ancient traditions, and especially the religious customs of negro paganism. This will be later a terrible barrier to civilization.

Among the Egbas, who form a sort of quasi-republic, the Ogboni are more powerful than the king. Nearly all the lawsuits are evoked by this tribunal.

The members recognize one another by different signs, principally by the manner of shaking hands. Death, and a cruel death, is the penalty for betraying the secrets of the society. The culprit as soon as he is judged is mysteriously condemned and shut in a narrow room; two holes are made in the wall a little distance apart, through which the victim's legs are drawn; the feet are firmly fastened to two stakes, then with a sword the flesh is slowly scraped from the thighs to the bone, and he thus dies in horrible torture.

The lodge where the meetings of the Ogboni are held is forbidden to those who are not members. From what I have been able to learn, this society is simply an institution similar to the secret societies of the pagan people of ancient times, where the members were initiated into the infamous mysteries of the good goddess. The divinity of the Ogboni is Ilé (the earth), one of the names of Odudua the great goddess of the blacks, who also has her rites and orgies in her official temple in the city of Ado.

Manes.

The blacks believe firmly in the immortality of the soul; hence their funerals are affairs of much more importance, and occasion much more expense, than a birth or a marriage. The disgrace of not having proper funeral ceremonies is such that often when a family has not the means necessary to defray the expense of a grand funeral the body of the deceased is wrapped in mats and preserved with aromatic plants in some secret place in the house; there is no mourning or wailing or weeping, but the family set to work to procure the required sum. When everything is ready, they suddenly break forth in loud sobbing and weeping as if the deceased had just rendered up his soul, and they proceed to bury the mummy. Others pawn their children to procure the means necessary for the funeral, and the children remain in slavery until they are bought again by paying the price stipulated.

The blacks believe that those who receive the honors of sepulture arrive safely in the country of the dead, called *orun rere* (the good heaven), which, according to general opinion, is situated directly under this world, so that the dead and the living may hold communication with one another. There the dead lead an existence very similar to our own, except that it is much sadder. Those who were slaves in this world remain so in the next; and those who were kings here are also kings hereafter. They have the same wants and the same likes that they had when living. Those who die without paying their debts cannot receive funeral honors unless the creditor consents to it. The body is placed on a wicker bier outside the city, and the relatives cannot bury it until they pay what he owed.

The bodies of great criminals are treated in the same way. They are placed on a bier outside the walls, and if the parents or relatives wish to bury them they have to buy the body.

When any one dies away from his country, the relatives do everything in their power to procure something that belonged to the deceased, no matter how small it may be,—a piece of his nails or of his clothing, or some of the hair,—and over these objects the funeral rites are performed, so great in their estimation is the necessity of funeral ceremonies; for those who do not receive funeral honors cannot go to the country of the dead, but are obliged to roam about in this world, exposed to the danger of being carried off by evil spirits, who cruelly maltreat them and cast them into the great fiery furnace called *orun-apadi* (the heaven of potsherds); that is, a place resembling the furnaces in which the blacks bake their pottery, a place covered with coal and the débris of the vases broken in the baking. The principal accompaniment of capital punishment is privation of burial. The greatest criminal has no fear of the next world if he is sure of funeral honors; for the negro has no conscience; with him all the wrong consists in being found out; he fears only temporal punishment, but above all being deprived of burial.

For information of a cherished relative and of his fate in the other world they apply to the fetich-priest, who takes a little child, bathes his face with lustral water, makes a sacrifice in a new vessel, and goes at midnight to the large park of the city or village, where he digs a hole in the ground, into which the child looks. Through this hole the child sees the dead under the earth, observes all that they do and say, and reports it to the fetich-priest. The priest, when he has learned all that he wishes to know, bathes the child's eyes with sacred water, and he immediately forgets all that he has seen and heard. This shows the skill with which the fetich-priests take advantage of the credulity of the blacks. Generally the dead are consulted by sacrifices and offerings made on their tombs.

Metempsychosis.

The blacks believe that the dead often return to this world and are born again. I saw a child whose mother dared not punish it, and submitted to all its caprices, because the fetich-priest declared on the day of its birth that it was the grandfather of the mother, who had returned to this world.

At Whydah an infant who had been born with teeth was thrown into the lagoon, the fetich-priests having declared that the child was the father of the reigning king, who had come back to earth. The king obliged his father to return to the dead.

While I was living in Porto-Novo I heard of a Nago killed in the war who returned, they said, and was born again of his own wife. The child bore on his forehead the mark of the ball that had killed his father; the mother affirmed that it was exactly in the same place where his father was struck.

Metamorphosis.

Of the numerous legends proving the negroes' belief in metamorphosis, I shall only cite two.

Bujé.

Formerly there lived a negress named Bujé, remarkable for her jet-black skin. She was sought in marriage by all the princes and men of wealth, but she treated all with equal disdain. One day one of the ugliest and most hideous of the negroes adroitly enticed her into his house and spread abroad the report that she had accepted him as her husband. Everybody believed it, and notwithstanding the belle's protestations to the contrary, she did not escape their raillery. She fled to the woods, and such was the violence of her chagrin, she was changed into a pretty little shrub which bears her name and which is used by the women to give their skin that ebony color which is considered the perfection of beauty.

Iyéwa.

A poor negress had two children whom she tenderly loved. She went every day to the forest to gather wood, which she sold to buy them food. One day they were all three lost in the forest; they walked a long time, but could not find the road. Overcome by hunger, thirst, and fatigue, they could go no farther and were obliged to rest. They stretched themselves on the ground crying and lamenting, and begging their mother for water. The mother having searched everywhere for the water returned to her children, whom she found almost dead. In her grief she addressed herself to Olorun Olodumaré, the all-powerful god, who heard her prayer. The mother, lying near her two sons, was changed into a large lagoon in which the two children quenched their thirst. Afterward they came and established themselves in this place and gave to the lagoon their mother's name, Odo-Iyéwa (the lagoon of Iyéwa, which is situated not far from Okiadan).

IV. ZOÖLATRY, OR ADORATION OF ANIMALS.

To the worship of the gods and the genii, the blacks join that of sacred animals. Each god has his favorite animal which is dedicated to him and serves him as messenger. All the time that they are in the service of the god the sacred animals are animated and directed by some inferior genius. For instance, the crocodile is consecrated to Osun, the wife of Chango. But all crocodiles are not sacred; only those designated by the fetich-priests as having the marks by which they are recognized as the official messengers of the goddess.

All animals may become sacred and be used as divine messengers. Some deformity or something unusual in an animal suffices to have it declared a fetich by the fetich-priests. This costs them nothing, but rather redounds to their profit.

Mepon and his Ox.

An ox which I presented to Mepon, the king of Porto-Novo, soon became his favorite. Every day the king gave him a small ration of acacia, and the animal never failed to come each day for his accustomed pittance. On market-days he went about among the crowds of negroes, never hurting anybody, and the king's favorite soon became the favorite of all. When Mepon died, the ox came as usual for his ration; but not finding his master, he began to bellow. The fetich-priests, understanding the cause of this, concluded that the genius of Mepon had passed into the animal. From that time it was forbidden to molest him; he was allowed to go wherever he pleased, and he never failed, especially on market-days, to take his usual walk. When he died, in 1883, the king had him wrapped in cloths and rendered him full funeral honors. According to custom, drums, gunguns, and every instrument capable of making a noise was brought into requisition, the blood of sacrifices and libations of palm-oil flowed in honor of the new fetich, while libations of rum were squandered to the great satisfaction of his adorers. Then the ox, followed by a cortege of fetich-priests and priestesses and the populace, was carried in great state on the shoulders of the negroes, and laid in the grave destined for his reception. He was besprinkled for the last time with the blood of the victims immolated at the tomb, and all was ended. The manes of Mepon ought to have been satisfied.

The multiplicity of the negro gods and goddesses shows us to what extent the idea of the divinity is adulterated among their disciples. Their adorers attribute to them marriage and posterity; they invest them with the tastes, the wants, and all the weaknesses and vices of humanity. There are wicked gods, drunkards, adulterous gods, liars, thieves, deformed and grotesque gods. There is no crime, debauch, or cruelty which their history does not contain. Thus the unfortunate negro, instead of finding in his religious beliefs a means of regeneration, sees therein examples and motives of perversion. The same corrupting influence is met with in their practices of worship, which is naturally in accordance with the divinities to whom they address themselves.

PART SECOND.

FETICH-PRIESTHOOD;

OR,

RELIGIOUS MINISTERS OF THE NEGROES OF GUINEA.

THERE are four orders of the fetich-priesthood, forming a hierarchy at the head of which is the king, who on the day of his consecration is initiated in all the mysteries of the negro sanctuary.

He then receives his new name. White is the official color of the vestments. His title as the religious chief is *Ekeji Oricha* (first after the fetiches). In Yorouba the chief of the Ogboni tries to take the place of the king in religious power. At Dahomey and Porto-Novo the king is all-powerful as long as he respects the national customs. He convokes councils of the fetich-priests on all extraordinary occasions, and is the judge who decides in all extreme cases.

In the reign of Mési, predecessor of Tofa, the present king of Porto-Novo, the fetich-priests wished to burn alive a young man who had by mistake killed a sacred serpent. According to custom, the sentence was to have been executed in a hut made of palm-branches covered with dried herbs prepared for this purpose. Everything was ready for the signal vengeance of the fetich; the fatal cabin was arranged, and the unfortunate negro, more dead than alive, was shut up in a fetich-hut awaiting his fate. On the eve of the execution, by some unknown means, but thanks no doubt to some of his relatives, the prisoner succeeded in escaping, and fled to Porto-Novo, where he offered his head to the king; that is, placed himself under his protection by constituting himself his slave. The next morning, to their great despair, the fetich-priests found the hut empty; but soon learning the fugitive's place of retreat, they all went like a band of furies to summon the king to deliver him up to them. The king, moved by the youth of the unfortunate culprit who had unintentionally killed the crawling god, wished to save him. He proposed to the fetich-priests to punish him and impose on him a heavy fine, but to remit the punishment of being burned alive, which he did not deserve. The fetich-priests would listen to nothing: justice must be done that the fetich might be revenged, and the death of the serpent must be expiated by that of the sacrilegious murderer.

Seeing that the king refused to deliver him up to them, they began to raise a frightful tumult: fetich priests and priestesses, wrapped in the most fantastic garb, some with their faces painted in red and white, others with feathers in their hair, or, according to their fancy or caprice, tatooed and smeared in such a manner as to render their appearance hideous, began to rush through the city crying vengeance and acting like creatures possessed. Then they returned to the palace, renewed their cries and vociferations, jostling the passers-by and striking terror everywhere, so that the market could not be held as usual.

The king, seeing his authority despised, summoned the Zangbeto (the police); then he notified all the fetich-priests to assemble the next morning in the great market-place to receive satisfaction. At an appointed hour the king had the gongs sounded, at which signal the

Zangbeto, who had secretly assembled in the palace during the night, rushed in a body on the fetich priests and priestesses who took flight; but a great number of them were seized, enchained, and sold as slaves to the profit of the Zangbeto. Some time afterward Mési died. It was generally supposed that he was poisoned; but he had made his authority as chief of the fetich-priests respected.

First Order.

After the king, the first sacerdotal order is the order of the Babalawo (the father who has the plate); they are also called interpreters of Ifa. Although Ifa is the third in rank of the superior gods, his priests form the first order of the hierarchy. They have two supreme chiefs: the first resides at Ifa, the sacred city of the blacks, and the other at Ika in Yorouba.

The office of Babalawo is to consult the fetiches, and to indicate what must be done to appease the gods and render them favorable, especially on important occasions, such as in war or during epidemics. They also watch over the worship of Ifa.

To this same order, but in an inferior rank, belong the Adahonche, whose office is the practice of medicine. Their medicaments are made with vegetables, to the preparation of which is added a thousand fantastical ceremonies to increase their value in the eyes of the negro. Besides Ifa, their divinities are Ochosin and Oroni, gods of medicine. To the first order are also attached the fetich-priests of Obatala and Odudua, who watch over the worship of these two divinities.

The insignia of the first order are white clothes, shaved head. a necklace of white pearls, and a cow's tail.

Second Order.

The second order is that of Onichango; that is, priests of Chango, the god of lightning. Their chief calls himself *Magba* (he who receives). He has twelve assistants:

the first calls himself *Oton* (the right arm); the second, *Osin* (the left arm); the third, *Eketu ;* the fourth, *Ekerin,* etc.

The chief and his assistants live at Oyo, very near Ikoso, where Chango descended alive into the earth and where his most honored sanctuary is situated. Their office is to watch over the worship of Chango. Their insignia is a bag, the emblem of pillage, which they wear in memory of the brigandage of their master.

To this order are attached all the fetich priests and priestesses of the lesser gods and goddesses, such as the god of the sea, of small-pox, of the lagoons, of the Niger, etc. The colors red and white designate the second order. The members shave the crown of the head, leaving the rest of the hair to grow, so that they look as if they wore a cap of black sheep-skin with the hairy side out. Sometimes they show considerable vanity in their head-dress: they wear a red and white striped cap, and arrange their hair in little braids after the manner of the women of certain countries.

Third Order.

The third order is that of Oricha Oko, the god of agriculture and of the men who have become fetiches. They devote themselves to white magic. Those who are addicted to black magic are not generally tolerated. They conceal themselves, and naturally have no particular insignia.

Consecration and Affiliation.

The priesthood of the false gods is hereditary in the family, the father being replaced by a member of the family. Others may be introduced into the corps of fetich-priests, but they have to pay dearly for the honor. The aspirants have to submit to an initiation of several years, which they must complete in a special college. The college of Chango for girls was established in a fetich-grove near our residence at Porto-Novo. Every morning before sunrise and every evening at sunset the aspirants were heard singing in choir, directed by an old fetich-priestess. Incommoded by being in our neighborhood, the college moved elsewhere.

The ceremonies of consecration of a fetich-priest last several days. The principal ones are the arrangement of the crinkled hair, which is completely shaved off of some and only from the crown of the head of others, the aspersion of lustral water, the imposition of the new name, of the new vestments, etc.

Besides this so-called consecration, there exists a sort of affiliation. Instead of the general jurisdiction of the fetich-priests, those who are affiliated are charged with the service of one god, but only in his special abode. The ceremonies of consecration and affiliation are very similar. We will only describe those of affiliation, which vary a little according to the fetich to which the candidate is consecrated, although identical in all the essential points. The candidate is generally a child, a boy or girl of from eight to fifteen years. As affiliation is expensive, very few can aspire to it. When the child's mother has saved money enough to purchase the happiness of seeing her son affiliated, she goes early in the morning to a fetich-priest, who with a band of his brothers goes in solemn procession to a fetich-grove. They begin by offering sacrifices to the gods to whom the aspirant for affiliation wishes to be consecrated. When the gods have breakfasted, the neophyte's head is shaved, he is stripped of his clothes, and bathed with a decoction of a hundred and one plants (there must be exactly this number); his loins are girt with a young palm-branch, and he accompanies the fetich-priests in procession around the sacred grove. During this procession the assistants remain prostrate with their faces to the ground.

When they re-enter the grove the neophyte is vested in his new clothes. Then the principal ceremony takes place. This is to ascertain if the fetich accepts the new priest proposed to him: this acceptance is an indispensable condition. He is consulted thus: The neophyte is seated in the fetich-chair, the priests bathe his head anew with the concoction of herbs and invoke the fetich. This ceremony is repeated three times; they at the same time dance and jump around the neophyte, making a deafening noise with drums and all sorts of old iron. Among the blacks nothing is done without music, and the more infernal the din and racket the more solemn the feast.

At the third invocation the neophyte begins to twitch, his whole body trembles, and his eyes become haggard; soon he becomes so violent that it is often necessary to hold or tie him to prevent his injuring himself or others. Then all the priests and all persons present hail the fetich with loud joyful cries of Oricha ô! ("It is the fetich.") Oricha gun ô! ("He is possessed by the fetich.") Finally, after several hours of frenzied tumult, the fetich retires from him, and he immediately returns to his senses. His violent frenzy suddenly ceases, and is followed by extreme prostration and lassitude. Some remain for quite a time as immovable as if dead. The fetich-priests and the assistants have the flesh of the victims cooked; then follows a great feast in the fetich-grove. When they have thoroughly fortified themselves, the person just affiliated is conducted with dancing and singing to a fetich-hut, where he must remain for seven days in the company of the god of whom he is supposed to have become the happy spouse: during this time he is forbidden to speak. When the appointed period has elapsed, the fetich-priests open his mouth, thus giving him permission to speak; they bestow upon him a new name, and the parents deposit shells at the foot of the fetich-idol, saying, "I buy back my son." They make some more sacrifices, and the fetich-priest learns at the initiation what things are allowed him and what are forbidden.

INITIATION OF AN AFRICAN FETICH-PRIEST.

These things vary according to the fetich, some, for example, are forbidden to eat mutton, others to drink palm-wine. Finally the fetich-priest teaches the neophyte the ceremonial to be observed in the worship of the fetich to which he is henceforth consecrated, and installs the symbol of the fetich in the neophyte's cabin. The person affiliated is supposed to belong to the family of the fetich-priest who initiates him; he cannot marry a member of this family, and he also becomes the heir of the fetich-priest should he die without children.

At the moment of the important test, if the neophyte is not possessed by the fetich, they conclude that he does not wish to accept him, and then there is no initiation. The pagan mother of one of our baptized children, about eight years old, wished to have her child initiated without the knowledge of the father, who was a Christian. The child would not consent to it, and resisted the caresses, threats, and blows of the mother and the fetich-priests. The priests forced him into the sacred seat of Chango and tried their incantations, but to no effect; the fetish did not come; and they were obliged to leave the child in peace.

The fetich-priests are neither loved nor esteemed; but they are terribly feared. Their person is sacred; and if a layman has the audacity to strike a fetich-priest he is severely punished. Lately the wife of one of our Christians reclaimed a sum which a neighboring fetich-priestess owed her. She refused to pay it. A quarrel ensued, and the priestess received a blow. She immediately raised a great cry; her sponsors the fetich-priests ran to her, and she told them of the sacrilege committed on her person. They all began to howl, and seizing the poor culprit, they put her in chains, beat her, and shut her up in a fetich-hut. Her husband and relatives had to pay a heavy fine for her release.

The higher order of fetich-priests can live on the revenues of their office, but the others, who are very numerous do not realize enough from their functions and are obliged to engage in divers trades.

In character the fetich-priest is a contemptible creature: deceitful, lazy, hypocritical, impure, and an arrant thief. He is generally very dirty, his clothes ridiculous and ragged; and those who steep their hands in human blood have a beastly, ferocious, and repellent appearance.

Beliefs of the Fetich-Priests.

The great or chief fetich-priests have a secret doctrine which differs greatly from the popular doctrine. In this secret doctrine they gradually initiate the priests of the lower ranks. These are the secrets of divers legal tests, such as Oncé, the test of the lagoon and that of Togo, and also the medical receipts, especially those for poisons. I do not believe there exist in the world more skilful poisoners. They preserve these receipts with great care, and much of the information contained in this work was only acquired by gaining the confidence of some old fetich-priest, principally by means of presents which not only strengthen friendship but are often powerful in loosening the tongue. As for the gods and goddesses, with their ridiculous legends, the great fetich-priests have no faith in them whatever. They despise the absurd beliefs and puerile practices which they foster in the people and even among the lower ranks of fetich-priests. They have no idea of the creation; and their idea of God, although vague and obscure, represents Him as the director and master of the universe. They believe in spirits, and are strengthened in this belief by the practices of magnetism and spiritualism. Nevertheless they have many superstitious customs which are no less ridiculous than those of the people.

The blacks are convinced that the different divinities inhabit, govern, and move the different parts of the universe in which they are incorporated, and which in obedience to their will produce good or ill and dispense the blessings or evils of nature. They conclude from this they must adore them, and offer them their vows and prayers. This worship rendered to their fetiches is absolute, for each god is considered to be a perfectly independent power in his domain; in his own sphere he may act according to his fancy. Chango, for example, thunders when he pleases; Elegba likewise perpetrates all the mischief that comes to his mind without consulting any one.

The fetich-priests have sometimes been compared to the saints whom Catholics invoke as mediators between God and man. This reproach may seem true to some Protestants, but it shows absolute ignorance of the subject. Nowhere among the blacks is there found a single example of a worship subordinate to a superior Being. They have not even the most remote idea of it.

The blacks not only adore the fetiches in the physical objects which they are supposed to inhabit and animate, as the sea, the streams, lagoons, animals, and trees, but they adore them also in the statues and the symbols which represent them and which are consecrated to them. They believe that the fetich-priests possess the art and power of intimately uniting the gods and genii to the material objects, and these objects once designated by religious ceremonies become like bodies animated by the gods with life and with sufficient power to predict the future, inflict maladies, excite the passions, do good or harm according to the will and pleasure of those who invoke them.

This, as may be seen, is far from resembling the homage rendered to the images of the saints, for Catholics have no idea of adoring animal matter. The blacks do not adore the stone, the tree, or the river, but the spirit which they believe dwells in them. During the first years of my sojourn on the Slave Coast, our neighbor the great fetich-priest of the thunder died, and all his fetiches were thrown out of the house as so many useless objects. I asked the negroes why they treated their gods thus; but they assured me that the gods were no longer in the fetiches. I inquired if the gods would not remain in the family under the care of one of the sons of the deceased, they replied they had gone with their servant the king. Therefore all the statues and other symbols of the gods, henceforth useless, had been thrown out.

Idols.

The statues and symbols of the gods are, according to the divinities they represent, statues of monsters, ridiculous objects, figures of birds, of reptiles or other animals; and these images, often shameful and scandalous, are in everybody's hands, in all the temples, houses, and public places, as well as along the roads. The indecent statue of Elegba is to be seen at the door of every house.

Temples.

A small cave, generally round in form, rarely square, built of potter's clay with a straw roof, painted on the inside in the color of the god to whom it is dedicated, very narrow and so low that the fetich-priest has to stoop very low in entering: such is the fetich-temple.

Grotesque statues and other symbols of the god, with dishes and earthen pots to receive the libations and offerings, all horribly smeared with palm-oil, blood, and chicken-feathers, form a mixture anything but agreeable to the sight, and still less so to the sense of smell, but worthy in every respect of the ceremonies of worship of the ragged fetich-priests and of the ignoble fetiches. What a contrast these uncleanly little huts present to the long spacious avenues shaded by magnificent trees which generally shelter them!

Besides these public temples most of the negroes have also at home their fetich-huts, sometimes kept quite clean, but always with the same style of idols, modelled on the ugliest type of negro, with thick lips, flat nose, receeding chin—a perfect face of an old monkey.

Groves.

Besides the temples in their beautiful shady squares, the blacks also dedicate to the worship of the false gods charming groves outside the city. Thither they all go in procession and revel in the dance in the open air under the cool shade of magnificent trees, the thick foliage of which shelters them from the burning rays of the tropical sun. In the middle of the sacred grove, in the midst of thick verdure, several bombax-trees spread their enormous trunks and grow to an immense height, superb giants stretching out their enormous branches, densely covered with leaves, like an immense parasol. In the centre of the grove are several small fetich-huts; a girdle of thorny trees encircles the grove itself, and leaves from the palm of Ifa scattered about indicate that the place is forbidden to the laity.

Talisman.

The blacks wear as ornaments, or as fetich-objects, necklaces, bracelets, and rings, the color of which indicates the god they serve. They have also either on their person or in their houses amulets or charms which they call "medicines." A piece of wood, a leaf, a bead, a tooth, an animal's claw, a bone, bird's feathers, all may serve as amulets; and the blacks have great faith in these things with which the fetich-priests furnish them, but by no means gratuitously.

An old fetich-priest of the Slave Coast was boasting of

the power of a talisman which he had made, saying that armed with his medicine he had no fear of anything; nothing could harm him—ball, sword, or knife. As I laughed at him and his panacea before the negroes who were present, he challenged me to harm him. It was easy to test his invulnerability. I sent him to get his talisman, and he soon returned, followed by a crowd of negroes hurrying to witness the encounter between the black and the white priest. The Brother doctor brought his lancet; the fetich-priest, with his famous medicine in his mouth, advanced swaggeringly and presented his arm without flinching to the Brother, who with a light touch of the knife made a slight gash. At sight of the blood which flowed, the poor fetich-priest remained glued to the spot, his ugly face, which could not blush, making frightful grimaces. All the negroes present shouted, assailing with an avalanche of raillery the poor fetich-priest, who, covered with confusion, hurried away with the Brother to the pharmacy to have his wound dressed. When he returned he had already recovered his assurance, and was prepared with a subterfuge. This medicine, made for the blacks, said he, has no efficacy against the white people. I immediately called a negro and asked him to bleed the old sorcerer on the other arm. This time he did not await the test, but fled, followed by the shouts of the spectators.

Sometimes merchandise may be seen deposited along the side of the most frequented roads with a sign designating the price of it; for example, a basket of bananas, on which are placed a certain number of shells, indicating the price of one banana. The vender leaves his merchandise there in perfect security; for beside it he has taken care to place a fetich-object, charged to guard it. No negro would have the audacity to touch it without leaving in its place the sum designated, for by so doing he would draw upon himself a terrible malediction. This custom is very advantageous for the vender as well as the purchaser.

There is a great difference between the talisman charms

or amulets and the animals and sacred trees. These are regarded as being possessed—that is to say, as the abode of a spirit; while the talismans are objects to which the gods have attached a special virtue which is thereafter inherent in the talisman, which of itself produces the effect, as brandy for example produces drunkenness. The fetich-priests claim to have the art of composing these talismans, and make them a source of great profit to themselves. Although these articles very often fail in their effects, the blacks have the greatest faith in them, and always find some good excuse for the inefficacy of the talisman.

The *Mégan*, the great executor of important works and first minister of Porto-Novo, possesses a collection of talismans of another kind. The walls of his house are entirely covered with human jaws to protect him against ghosts. Every time he executes a criminal or immolates a human victim, the old executioner keeps the jaw, which he hangs up in his house. Without this precaution this just man's sleep would be disturbed by the dead coming, weeping and wailing, and knocking at his door.

Ceremonies of Worship.

Sacrifices are the most essential part of the worship. On all occasions, even the most unimportant, nothing is ever done without consulting the gods by immolating victims to them.

Every five days the fetich-hut is swept, and a supply of fresh water and provisions, which are invariably sprinkled with palm-oil, are placed before the idol. Similar offerings, more or less generous, may be renewed each day, according to the devotion of the negro who seeks to ingratiate himself in the favor of the fetich.

On solemn occasions the fetich-priest is consulted and directs the sacrifice. He designates the victim, which must be pure, to offer to the fetich, for each fetich has his pure and impure animals. While the blacks kneel, he presents to the idol the suppliant's request thus: "Behold the victim they offer thee; listen to their prayer; may they be in peace," etc. The more eloquent the fetich-priest is—the sweeter his tongue, as the negroes say—the more he is in demand. He then immolates the victim and pours a little of the blood on the idol; the head and intestines are placed in an earthen dish, which they deposit in front of the fetich-hut, with an accompaniment of a sort of infernal music impossible to describe. The fetich-priestesses, under the direction of some of the fetich-priests, jump about like a band of furies, executing a dance around the musicians with lascivious and ridiculous movements. From time to time they stir up the sacred fire by copious libations of brandy, and song and dance succeed each other with a frantic frenzy of which it is impossible to give an idea. The blacks come in great numbers from every direction and crowd around the musicians and dancers; even the children dance with delight.

After a day and night of this tumult of mad folly, all stop to rest and sleep. When they have danced for a time in one place, they go and begin the same scene in another; and this lasts four, six, nine days, and sometimes longer. There are several feasts of this kind during the year, but the principal one, which is called Odun (the year), is celebrated about the first of October.

Human Sacrifices.

Ugun, the terrible god of war, is not satisfied with the blood of animals, but like the dreaded Elegba must be appeased with human blood. In wars and public calamities, human victims alone can satisfy the angry gods.

Human sacrifices are generally offered in the night. No one is allowed to leave the house. "The night is bad," the blacks say. The sound of the drum and the dismal chants of the fetich-priests alone indicate that human blood is about to be shed before the idols. The victim is gagged, and his head is cut off in such a way as to allow the blood to gush forth on the idol; then the body is dragged along the ground and thrown into a ditch or into the bushes. But first the fetich-priest opens the breast and takes out the heart, which he keeps and has dried to make talismans, and also to give courage to the combatants in war. The heart when dried is reduced to powder and mixed with brandy, a ration of which is given by each chief to his men.

If the sacrifice is offered to the lagoon or the sea, the body is thrown into the water. For the evil spirits like Elegba the body is opened, the entrails placed before the idol, and the body suspended in front of the fetich, where it is left to putrify and fall to pieces. I have sometimes seen these bodies on the roadside, and have been obliged to go out of my way to escape the infectious odor they exhale.

These human sacrifices are offered for different reasons. One day, for example, a prince of the forest being ill consulted Ifa. The answer was that the illness came from an angry spirit. When consulted again, Ifa replied that the illness would not cease until a human victim had been offered to the spirit, and the victim was immolated.

Another prince, at war with Porto-Novo, seeing that his soldiers lost courage, had recourse to his fetiches, who recommended to him a powerful charm. To compose this, a little child was carried off, while the mother, a young slave, went to draw water. The child was thrown alive into a mortar and pounded to death, and the fetich-priest made charms of it for the prince and his soldiers.

Birth.

At the birth of a child, a fetich-priestess takes care of the mother and the child. On the ninth day (for a boy), or on the seventh day (for a girl), a fetich-priest of Ifa is called, who kills a hen and a cock in honor of Ifa, and of the good genius of the head of the child. The entrails of the fowl are sprinkled with palm-oil, and are carried as usual to Elegba to prevent his coming to disturb the ceremony. This done, they take the fresh water which is renewed every five days and placed before the fetiches, and throw it on the roof of the cabin, and the mother, carrying her infant, comes out and passes with the child three times under the water which falls from the roof.

The fetich-priest then makes the lustral water, which he prepares with snails and vegetable butter. If the circumstances of the birth did not indicate the fetich who took the child under his protection at his coming into the world, they consult Ifa, who makes it known. After this the fetich-priest bathes the forehead of the child with lustral water, repeating three times the name the parents wish to give it; then he makes the child's feet touch the earth.

They sweep and clean the cabin, take away the fire and ashes, and when all this is done they make a fresh fire. A sacrifice to Ifa, followed by a feast, terminates the ceremony.

Forty days after this the mother shaves her head, makes her toilet, and pays a visit to the fetich-priestess who attended her, and offers with her a small sacrifice to the fetich of the child. She goes afterwards to visit her relatives, and after that day she returns to her usual occupations.

Marriage.

Before marriage the first thing to be done is to consult Ifa to know if the marriage may take place, and if it will be happy. If the answer is in the affirmative, the ceremony is decided upon. On the appointed day sacrifice is again offered to Ifa, and the two who are betrothed con-

A FETICH-DOCTOR PRESENTING HIS DIPLOMA TO REV. FATHER BAUDIN.

sume at the wedding repast the viands offered to the idol. They eat, drink, and amuse themselves until midnight. The bride is then conducted by her companions to the nuptial cabin, where the husband soon follows, and all the others withdraw. If the bride is not found worthy of her black husband, she is punished and sent away; he must then return her dower and the presents, at least until the affair can be arranged. If, on the contrary, she is agreeable, the mother is complimented and receives a present of white porcelain shells. Finally the old women present the bride with kitchen utensils. The mother explains to her son-in-law the character of her daughter, how and on what occasions he must correct her, and, in general, the manner in which he must treat her.

Funerals.

The moment a negro dies, the women, young and old, rush from the death-chamber out into the court-yard, utternig piercing cries and lamentations. Some, with their hands clasped over their heads, weep, howl, and stamp their feet; others run from side to side, stop suddenly, clasp their hands on their heads, and begin again to jump frantically—in a word, manifest the most violent despair. The neighbors run to learn the cause of all this tumult, and the racket is then only increased, the clamors redoubled. The women will not be consoled; they wish to die; some fling themselves on the ground, others seem to wish to break their heads against the wall. The neighbors restrain them and do their best to console them, but in vain. The children, bewildered by all these performances which they do not comprehend, begin to cry on the backs of the negresses who rush about in every direction as if crazy.

After this first tempest there comes a moment of calm. They tell the neighbors how the deceased died; that they never supposed the end was so near, and they did everything to ward off so great a misfortune, etc.

They then go to notify the relatives, who hasten to return with the neighbors, and the scene just described is repeated: renewed tumult and groaning, fresh floods of tears, new clamors. The relatives end by consoling them all, make the women retire to a room where they can rest, weep at their leisure over the dead, and take care of their children.

The eldest son consults with the women about the funeral and the solemnity they wish to give it. He sends for a priest of Ifa, a *babalawo*, who, having immolated pigeons and hens, consults his fetich to know if there is any need to appease the gods to keep off the evil spirits and other dangers which might menace the deceased or his family. If Ifa replies in the affirmative, the fetich-priest offers in sacrifice a goat, of which he opens the stomach, sprinkles it with palm-oil, places it all in a basket or in a broken vase, and has it deposited outside the city, at a place where three roads cross each other, so that the evil genii and other imps can take whichever road suits them in their flight.

The *babalawo* then makes the lustral water in an earthen pot with the slime of large snails; he sprinkles the death-chamber and the assistants, using for this purpose a fetich palm-branch, and prays the deceased to depart gently and quietly, saying at the same time, "May God show thee the good road; mayest thou meet nothing evil in thy way;" and other prayers.

While some of the relatives are having the hens, snails, and other food cooked, the others begin the toilet of the deceased. He is bathed from head to foot with a decoction of aromatic plants, and then with brandy if the deceased is rich enough to afford it. His hair is shaved off and tied up in a white cloth and buried behind the house. He is dressed in a *chokoto*, a kind of drawers which the blacks wear as trousers; the head is covered with a cap; the hands are laid on the breast; the thumbs are tied together, as well as the great toes; then he is decked with necklaces, bracelets, and rings. If it is a woman, she is painted with a reddish powder, made of dyed wood mixed with vegetable butter and other fragrant substances.

The body is then wrapped in a great many cloths; each relative having brought one for this purpose, so that there are sometimes as many as forty.

The body, which now forms a large bundle, is exposed on a funeral mat at the door of the death-chamber, where it must remain three days. The daughters or sisters of the deceased seat themselves on each side with fans to keep off the flies.

During this time a grave is dug in the cabin-floor. It is a deep cut at the bottom of which is a subterranean gallery in the form of a cave, contrived in such a way that the head of the dead when buried is outside of the wall, under the veranda, and the feet inside of the cabin. A coffin is also made of rough planks of native wood.

Meanwhile the living are not forgotten. A funeral is a great feast: they must above all drown their sorrow. So the evenings are spent in eating, drinking, dancing, and singing for the dead. Drums and iron instruments are beaten, and guns fired: this is the obligatory part of the ceremonial. Frequent visits to the gourds filled with palm-wine and rum keep up the enthusiasm.

At the beginning of the feast, the widows and daughters of the deceased are conducted to a room near by, where they are obliged to remain for three days. Their tears, cries, and shrieks mingle with the noise of the drums, songs, and the firing of guns. Finally the relatives, well fortified, go to console them and beg them to eat. They at first refuse: " How can we eat when our dear one is no longer with us? No, we wish to die with him; we care no longer for food." The relatives beg and implore them. They end by yielding, and consent to take something to sustain their miserable existence. At sight of gourds filled with viands sprinkled with palm-oil and strong spices their sadness is dispelled a little; bottles of palm-wine and rum are also secretly supplied to them. The eldest son has every interest in gaining their good graces, for, according to a degrading custom, at the death of the father the women are divided among the sons; no one, however, is allowed to take his own mother.

The night and two following days are spent in orgies with intervals of repose. Every day in the morning, at noon, and in the evening the lamentations of the women are heard soliciting the same consolations. The third day, after a plentiful repast, a band of negroes place upon their heads the bier on which the body rests, covered anew with a handsome cloth, and run with it through the city, while the others throw shells to the crowd following them and jostling one another to gather them up. The bearers skip, jump, and make a thousand extravagant contortions while singing the praises and celebrating the wealth of the deceased.

They return in the evening and proceed to bury him. The body is put in the coffin with shells, brandy, and other articles. They secretly take away the cloths, each relative taking his own, which he carefully hides. The coffin is lowered into the grave, covered with mats so that the earth may not touch it, and sprinkled with the blood of a he-goat which is immolated at the tomb as a sacrifice of expiation. The negroes then throw shells and handfuls of earth into the grave, and take leave of their dead, saying, " Safe journey! May God grant thee to arrive in peace! Mayest thou not stray either to the right or left!" vying with one another as to who shall express the greatest number of good wishes.

In some places they leave the head of the grave open, and afterward take away the head of the deceased, which they place in a fetich-cabin, where they make offerings to it.

When the grave is entirely filled in, the orgies recommence and last all that night and the next morning. After having slept, in the middle of the day the band of negroes run again through the city as if seeking the dead. A choir sings, *Baba wa l'a nwa, awa à rii.* ("We seek our father and find him not.") The others answer, *O rè ile; ó rè ile re.* ("He has gone to his house; he has gone to his home.")

The feast and racket continue until the evening of the next day. Then they collect the bones of the victims immolated and eaten and hang them on the wall over the grave. The more bones there are the more solemn is the funeral. A band of negroes armed with guns, and followed by other negroes, carry the mat, gourd, shells, brandy, and other treasures of the deceased, all of which they shatter with shots and burn in a fetich-grove outside the city, to signify to the dead that he must depart forever, for there is no longer anything for him in this world.

During this time the young men kill a hen, the feathers of which they scatter as they go; they have it cooked and eat it on the roadside near the grove. This is what they call *Adic-Irana* ("the hen which buys the road"). It is supposed to precede the dead on his journey and show him the way.

On his arrival at the gate of the other world, he pays to pass through, and thus happily reaches the country of the dead, which is called *Orun rére*.

During the funeral, the relatives, in token of mourning, neither bathe nor comb their hair. The last day they shave their heads and go to visit the relatives and friends who have come to console them; then the mourning continues for from three to twelve months, according to the localities, and consists with the blacks in leaving their woolly hair uncombed.

From time to time the blacks make libations and offerings on the tomb, they offer sacrifices, and by means of this kind consult the dead on the most important occasions.

Their faith in the immortality of the soul, and in the intercourse which the dead may have with the living, is evident from the ceremonies which I have just described; and the stories they relate in the evening, seated on mats in the moonlight, enjoying the fresh air, also evince this belief.

The following narrative proves the negro's belief in the influence of the dead:

One day a woman deposited with another negress in the presence of witnesses a coral necklace. Then she went to get salt, a great way off, on the sea-shore, where the people on the coast prepare it by vaporizing the salt water, first in the sun, and then on the fire in earthen pots. The woman who received the necklace carefully hid it in a hole which she made in the wall of her cabin, and which she plastered over again in such a way that it was impossible to discover where the hole had been made. It happened that she died suddenly without being able to make known to her two sons the hiding-place. They having rendered the last honors to their mother, searched everywhere for the necklace, but could not find it.

The negress returned from her journey and claimed her necklace. The two sons related to her the whole affair; but she would not believe them, and accused them of theft before the king, who heard them and also refused to believe their story. The youngest was put in prison, and the house was to be confiscated if in eleven days the necklace was not returned.

The eldest son, not knowing what to do, addressed himself to the grand priest of Ifa, begging him to assist him. Touched by the young man's grief, the fetich-priest consulted Ifa, who replied that he must go to the country of the dead and ask his mother where she had put the necklace. "Let the young man," said he, "offer this evening a black sheep to the dead near the sacred grove outside the walls; let him bathe his eyes with lustral water and follow the first dead person he will see pass, and he will reach the road to the dead. By paying the passage-money the gate-keeper will let him enter. But let him take care not to touch the dead, for if he does he will never see again the land of the living. When he returns to the sacred grove, he will bathe his eyes again with lustral water and offer a victim to the gods who have allowed him to visit the country of the dead without dying."

The young man accomplished all that had been prescribed, and happily reached the end of his journey. The first person he met was his mother. She was going dejectedly toward a fountain; the other dead were seated here and there, or walked about silently. Seeing his mother, he cried:

"Iyà!" (Mother.)

She raised her head, recognized him, and came to meet him.

"What! is it thou, my son? Why hast thou come down to the abode of the dead?"

"My brother is in irons, and our house is to be sold if our neighbor's necklace is not restored to her. The great Ifa has allowed me to come among the dead to ask you where you have put it."

The mother told him where it was hidden. The young man having obtained what he had so earnestly wished for, forgot the recommendation of Ifa, and was about to throw himself at his mother's feet; but she drew back.

"Do not touch me, my son, lest the road to the living be closed against thee forever. Return and deliver thy brother; offer to thy mother victims and make to her offerings, for in this place she has great need of them."

She withdrew and disappeared.

The young man returned, went to the sacred grove

and offered to Ifa the promised victims. He found the necklace, and was fêted throughout the city. He did not forget his mother. Every five days he replenished the water on her tomb, and from time to time presented her offerings and victims.

The blacks believe that the land of the dead is very similar to this in which we live, but much sadder. The dead have in the next life the same position they had in this; those who were kings are so still, and those who were slaves remain slaves. They have the same pleasures, the same habits, the same needs. Hence it is considered a duty, an act of filial piety, to offer them libations and sacrifices. The kings, chiefs, and persons of wealth must be furnished with a retinue of women, and slaves to keep up the dignity of their position and secure them the comforts suitable to their rank. At great intervals messengers are sent to inform the dead of what is transpiring in this world, to interest them in the welfare of the country, and to obtain their advice on important occasions. They become enraged with the living if they do not liberally satisfy their wants and desires, each one according to his resources and position. But they are on the contrary pleased when they immolate on their tombs the enemies with whom they formerly fought.

These ideas and beliefs are the real cause of the human sacrifices which every year imbrue in blood these unfortunate countries of the blacks, as well as of the brigandage and continual wars necessary to procure the victims. At the death of the kings and chiefs, victims are immolated beside the grave, and their blood gushes forth on the coffin; the women and slaves are massacred that they may accompany the dead, to serve them in the other world. From time to time they send them other women, new servants, and often even messengers to acquaint them with what takes place on the earth.

One day the king of Dahomey had thus dispatched several couriers to his predecessors, when he remembered

some insignificant detail of his commissions that had escaped his mind. A poor old woman was passing, carrying on her head a pitcher of water. The king called her and gave her his message. The poor wretch, trembling all over, begged and implored for mercy.

"I have done nothing wrong," she said.

"I know that," replied the king, "but I am sending you to my father; go at once."

Resistance was in vain. The poor creature knelt down, drank half a bottle of brandy, and the Méhu cut off her head.

The women, slaves, and messengers who are sent to serve the dead are beheaded. But their enemies, especially the chiefs and fetich-priests whom they have conquered expire after insults and taunts, in the midst of frightful tortures; then a barbarous scene takes place, too hideous and repulsive to describe.

At Porto-Novo I have attended royal funerals which have lasted nine days and cost the lives of numerous victims. One victim was skinned, and of the skin a drum was made to be used in the ceremony. In the market-place, around the bodies, the negroes drank brandy at will, danced and gave themselves up to all kinds of amusements.

Dahomey especially has acquired a sad notoriety, from the massacre of human victims which accompanies the annual feast, called "the feast of customs." Every year the army of Dahomey takes to the field to secure among the neighboring tribes their supply of victims, and also of slaves which they sell to buy brandy, powder, and the things that they send to the dead, as well as the rewards which the king distributes to his soldiers and his people, for whom the occasion of the yearly sacrifices is a time of feasting and rejoicing.

This system of annual brigandage has made a vast desert around Dahomey. The latest news from Guinea brings us tidings of the destruction of Ikétou, the only important city that remained in the western part of the Nagos countries, formerly thickly populated, now only a desert abandoned to the wild beasts.

This year the Dahomeans spread the report that they had been beaten by the Mahis. The Ikétous, supposing their mortal enemies far away from them, thought only of enjoying themselves and took no precautions. Meanwhile, the Dahomeans, following their usual tactics, glided like serpents through the trees and underbrush of the forest, creeping stealthily and silently until they reached the walls of the city, when each man went to his post, and with gun in hand awaited the signal.

The first cock-crow was the signal. Then the Dahomeans scaled the walls, even before the inhabitants, surprised in profound slumber, could recollect themselves. Those who tried to escape were seized and garroted; those who attempted to resist were massacred on the spot; men, women, and children were bound together in groups outside the city.

The pillage was soon accomplished, for the blacks are not rich. The city was then set on fire, and the sick, the infirm, and nursing children were thrown into the flames. The captives, chained together in long lines, were led to Abomey to be divided as slaves between the king, the chiefs, and the soldiers, except those who were put aside to serve as victims in sacrifices.

This is merely one instance of the bloody tragedies which are repeated every year.

CONCLUSION.

The study of this subject has shown us the deep religious sentiment of the blacks. On all important occasions in life the fetich intervenes. This idea the negro carries everywhere with him; there is no such thing, then, as indifference in matters of religion.

But how perverted is this religious sentiment which the negro has inherent in his nature in common with all men! First the Divinity loses his principal attribute, goodness. Olorun is not mischievous, hence they do not trouble themselves about him. It is the evil genius Elegba who is never forgotten; offerings are made to the other gods and demi-gods only to escape their vengeance. Fetichism is the worship of cruelty and also of vice. The most immoral negro is as good as his fetich.

A strange confusion of good and evil, an incoherent mixture of doctrines, fetichism presents the most clearly defined spiritualism and the most repugnant materialism: monotheism in Olorun, the supreme god; polytheism in all the lesser gods; the government of the world by a superior but wicked power; the immortality of the soul; future life, but without reward of virtue or punishment of vice, consequently no conscience; respect for the dead, but a respect stained by the human sacrifices by which it is expressed. In a word, it is a complete perversion of religion, which, instead of elevating man to God, serves only to debase him.

Yet there is a lower and still more vile being than the black fetichist, and that is the fetichist turned Mussulman. To his former brutishness and superstitions he adds two new vices: fanaticism and pride, two great obstacles to Christianity. The pagan negro becomes converted and confides his children to the missionaries; the Mussulman negro is unapproachable.

The rapid progress of Mahometanism in these countries alarms all the friends of Africa, all those who take an interest in these unfortunate people, and follow the march of events.

What will save them from this new danger? Colonization by white people is impossible because of the fevers and of the climate; moreover, fetichism with its human sacrifices has not disappeared from the vicinity of the business houses established for so many centuries on the coast. It is only in the circle where the influence of the Christian missions is felt that fetichism has lost credit, and this fact indicates to us the remedy of the evils of which we have only been able to give a faint idea. In Catholic evangelization and the charity of Christian nations rests the only hope of the salvation of the black fetichist.

FIRST CATHOLIC MISSION IN LAGOS (WEST COAST OF AFRICA) DIRECTED BY THE FATHERS OF THE SOCIETY OF AFRICAN MISSIONS.

REPORT OF REV. FATHER A. PLANQUE, SUPERIOR-GENERAL OF THE SOCIETY OF AFRICAN MISSIONS, AT THE GENERAL ASSEMBLY OF 1884.

Early in the year 1856 a young bishop, Mgr. de Marion Brésillac, came to Rome, after twelve years' missionary labors in the East Indies. Desiring to devote the remainder of his life to the conversion of the most abandoned tribes of Africa, he opened his apostolic heart to the eminent Prefect of the Propaganda, and proposed to go with several priests to establish a mission on the Slave Coast. Feeling that the European nations owed a great debt to these people, with whom they traded for so long a time for the benefit of their inter-tropical colonies, he wished to bestow upon them the true liberty of the Gospel. His project was favorably received at Rome, but it was not thought possible that a bishop and a few priests could successfully undertake a mission of this kind. He was advised to form first a society of priests, which would be a preparatory seminary and a reserve corps. This was a great undertaking, but the voice of the Holy See is the voice of God.

Mgr. de Marion Brésillac did not found a religious order, but simply a society of secular priests, bound by the general rules of the Holy See and by a community of interests. He established the centre at Lyons, under the title of Seminary of the African Missions, and on the 8th of December, 1856, he consecrated to Our Lady of Fouvière the first-fruits of the Society.

The foundation of a society is always difficult, but it becomes still more so when the enterprise must be undertaken in a distant country and there seems no tangible means of arriving at the desired results. The Bishop experienced at first many disappointments without, however, losing courage.

The Propaganda prudently made inquiry as to now the missionaries would be received in these countries of terrible repute. When informed that the missionaries would be murdered by these barbarous people as soon as they arrived, the Propaganda, not wishing to send them to certain death, created the Apostolic Vicarate of Sierra-Leone and confided it to Mgr. de Marion Brésillac, little thinking that this was sending him to a martyrdom less glorious, less desirable, and also less fruitful than that of blood.

Mgr. de Brésillac sent out two priests in the month of December, 1858, and came himself to Freetown in May, 1859, with another priest and a lay-brother. He found the city a prey to a nameless epidemic—the most violent it had ever experienced. The captain of the vessel tried to dissuade them from landing, but no persuasion could deter them from going to their perilous post. The Bishop saw two of his priests and the lay-brother perish before his eyes, and when he and his Vicar-General were attacked by the disease, they had already buried nearly all their Christians. Finally, within an interval of two days, they in their turn succumbed about the end of June.

Mgr. de Marion Brésillac's last thoughts were of his work, so cruelly stricken in its birth; and which, humanly speaking, seemed crushed at the very outset. But the founder and his four companions were not lost to the work; they watched over it from heaven.

The young scholastics in the Seminary of African Missions remained firm in their vocation, and when Pius IX. learned of the death of Mgr. de Marion Brésillac, and that his children were resolved to continue what he had begun, he sent them his special blessing. The Propaganda in a letter to the Superior expressed its pleasure and admiration at seeing that, far from being discouraged by the trial they had experienced, the seminarists seemed inflamed with greater ardor. And having obtained more accurate information of the state of Dahomey, it resolved to accede to the first request and constant desire of Mgr. de Marion Brésillac. The Slave Coast was consequently erected into an Apostolic Vicariate and confided to the Society of African Missions.

God did not abandon those who put their trust in Him, and vocations were not wanting. On the 5th of January, 1861, three missionaries embarked for Dahomey. The state of this country may be briefly described. Satan reigns supreme over these unhappy people; spiritual degradation has here reached its furthest limits; no idea of Divine truth exists among them; the darkness of gross fetichism envelops the land. Serpents, the thunder, hideous animals and still more hideous idols are the gods they adore. The principal feature of their barbarous worship is human sacrifice. The number of victims is unlimited, and they are immolated with revolting cruelty. Blood must sprinkle everything; trophies of death decorate every place; nothing of importance takes place without the shedding of blood. Every day the earth must be watered by the blood of some victim.

In the midst of this depravity God had prepared a way for the introduction of Christianity. The old French and Portuguese settlements along the coast had left some remnants of Christianity, and among the thousands of negroes which the slave trade had transported to other countries some had returned, principally from Brazil, where they had been baptized. If they did not bring back with them much knowledge, they had at least a great love and esteem for Christianity. Several generations of slaves, in returning to their homes, thus sowed the good seed. Moreover, the natives of these countries consider our religion superior to theirs, and the God of the white man much greater than their fetiches. The aged and men of mature years content themselves with this speculative esteem, for it is difficult at that age to divest themselves of materialism. But they willingly confide their children to the missionaries, and these children become excellent Christians.

It is not my intention to give a detailed account of the twenty-two years' apostolic labors of the African missionaries, in the early part of which they experienced many sad trials. Much good was effected, but the fevers and other maladies peculiar to this trying climate soon decimated the ranks of these zealous laborers. It took years to acquire the experience necessary to resist the severity of the climate. But the greatest difficulties were gradually lessened. The grain of mustard-seed has not yet become a great tree, but it has put forth promising branches, which I will briefly describe.

Until within a few years the Society of African Missions had but one station. Now the Holy See has entrusted to their charge four Apostolic Prefectures, which include the Coast of Benin, Dahomey; the Slave Coast; the Ivory Coast; and part of the Egyptian Delta.

Apostolic Vicariate of the Coast of Benin.

This Apostolic Vicariate was for a long time the only mission we had. The great results effected here show how much greater fruits might be reaped if we had the means necessary to send a larger number of missionaries to found new stations. Nine hundred and forty children now attend our schools, and these schools are the true centre of our Apostolate. All the children under our care become Christians, and form a nucleus which increases daily. They marry in accordance with the law of God and the Church, and even the old polygamists praise and admire this young generation who re-establish the unity of the family.

Lagos, the centre of the English colonies, is a city of no less than fifty thousand inhabitants, has an extensive commerce, and is often called the Liverpool of Africa. Our Gothic church is its finest ornament. All contributed to its erection; the government by a grant of land and the work of its people, the European commercial houses by liberal subscriptions, and the laborers by gratuitous work in their different trades. We have a school at Lagos for boys and one for girls, besides a higher school for the training of teachers. This training-school will become a nursery of most useful auxiliaries to the missionaries, by furnishing them with the staff necessary to the founding of new Catholic schools in the cities where missions are already established, as well as in the neighboring towns. There is very little idea in civilized countries of the density of the populations on the coasts of Guinea. The cities and large towns teem with inhabitants, among whom a line of schools might be advantageously established within certain distances. Heretofore we have not been able to accomplish this for want of the necessary teachers. Our high-school at Lagos now begins to obviate this difficulty, by placing at our disposal catechists and instructors for the schools. These teachers marry the young Christian girls educated by the Sisters, and the schools are under the supervision of a missionary who frequently visits them. In 1883 the first establishment of this kind was opened. It was followed by many others as soon as we could defray the first expenses of installation. Thus with a comparatively small number of missionaries we have been able to reach in our ministrations great numbers of the population. There are at present four other schools in prospect. In the most important centres we establish schools for girls and resident missionaries, and we rely on these central schools to establish a series of schools. It is not the work of a day, but the missionary is accustomed to be patient. He comes not to this land as an explorer, but to dwell in this new country of his adoption and to disseminate the great Christian principles which elevate the people.

The high-school at Porto-Novo will have still further influence by furnishing the colonies with honest and intelligent employés. Already the children from our ele mentary schools are sought for, and preferred to all others for ordinary service, because of their honesty. The government officials and the commercial houses ask us for young Catholics. This is a precious testimony to the efforts of the missionaries.

In our village schools, while imparting solid Christian instruction, we endeavor to instil in our pupils a love of cultivating the soil and of everything that pertains to agriculture.

Porto-Novo, the capital of a small kingdom of the same name, has a population of some thirty or forty thousand inhabitants. The people are principally fetich-worshippers, but Mahometanism already numbers many adherents. The missionaries have always exercised a great influence here, yet fear of the fetich-priests and the rivalry of various sects retarded for a long time the progress of our schools. They have now developed to such an extent that our accommodations are totally insufficient for the numbers of children who present themselves.

Aggéra and Aguégué, villages of several thousand inhabitants, have for a long time desired schools, and we hope soon to be able to satisfy them.

Porto-Novo has a beautiful church which is a marvel for this country. It was blessed in 1878. The king, Tofa, assisted at the ceremony with great pomp, and through respect for the great God of the white man dispensed with the usual ceremonial which requires the attendance of a certain number of his wives to fan him, carry his parasol over his head, etc. The conduct of the king on this occasion raised the missionaries very much in the opinion of his black subjects. They all said, "The God of the white man is much greater than ours, and His priests are superior to our fetich-priests." On Sundays, while the Christians are in the church, the pagans, fearing through respect to enter, crowd around the doors and windows, silent and recollected during the Mass. On feast-days and during the Novena for the Immaculate

Conception crowds surround the entrances to the church and join in the joy of the Christians.

The King of Porto-Novo has just placed himself under the protection of France; his kingdom is one of the most fertile and the best situated in Guinea, and is the principal commercial port for the exportation of palm-oil. France may acquire a great influence here without sending an army or without enormous expenditure. A few patriotic Christians residing near the King Tofa would give an influence to the French protection incomparable to anything on the coasts.

The Catholic missionaries take no part in the politics of the country; they are interested only in the Christian civilization of the people.

I wish before making any further statements to do full justice to our neighbors, who are ever ready wherever they find the missionaries to aid and assist them.

St. Joseph's, at Tokpo, is an experimental farm which has the double object of furnishing resources to the Mission and of gradually forming at a distance from the corrupting influence of the large cities a Christian village devoted to agriculture. The government at Lagos has ceded to us a tongue of land near Badagry, between the lagoon and the sea; it is fourteen kilometres long and about twelve hundred metres wide. It comprises meadows and forest of divers fragrant shrubs with thick underbrush. We have tried the cultivation of various crops, principally cocoa-trees. The first seeds sown suffered much from the rabbits and squirrels. Ounces, panthers, leopards, wild-boars, porcupines, monkeys, boas, and serpents of all kinds made merciless war on our flocks as well as our crops. But the gradual clearing away of the bushes in which they lived has driven them further and further away and finally banished them so far that with a little vigilance we have nothing more to fear from their ravages.

Some of the trees have already borne fruit, and the

cows, goats, and sheep of the flocks have rapidly increased to the great delight and encouragement of our laborers.

The farm is worked by two missionaries, a few orphan children, and a few that we have redeemed from slavery, who are learning to cultivate the soil and are at the same time taught the truths of Christianity.

When the land is sufficiently cleared, we intend to settle on it several reliable Christian families who will work the farm under our direction, and form a village which we hope will serve as a model for many others.

Abeokouta.

A canoe-journey of five or six days up the Ugun brings us to the city of Abeokouta, which is surrounded by a high wall about thirty-five miles in circumference. This city was built by the numerous tribes of Egbas of Yorouba, who, to escape the incursions of the King of Dahomey, established themselves here under the shelter of massive rocks. The city is divided in seven quarters, each of which has its respective king, but there is but one chief of war. It was through the instrumentality of this chief that Divine Providence established here our mission.

For a long time the evangelization of Abeokouta was the object of our missionaries' greatest desires. Two priests went there in 1880. Their reputation as men of God had preceded them. Several Christians who went from Lagos to live in this great city had often spoken of them; even some of the natives of Abeokouta, who had been in Lagos on business, had observed the good done there by the priests. Consequently they were very favorably received by the chief of war, Ogudipé, who being an intelligent, sensible man, and knowing the missionaries, appreciated what a boon their presence would be to the city of Abeokouta. He procured for them from one of the kings of the country two grants of land: one for themselves, and the other for the white women who never marry and who come from cold countries to teach the children. They thus designate the Sisters of the Society.

Ogudipé smoothed away all obstacles, and two missionaries with a catechist were able to install themselves in the great city, where they have built a small house and a school. But here are only two white missionaries to minister to one hundred and fifty thousand blacks, for Abeokouta contains this number of inhabitants at least, as nearly as can be ascertained in a country that has no statistics. They already have quite a numerous flock, among whom may be seen several influential men. A house for the Sisters is being built, and a large church is very necessary here. Thus is the very heart of the vast continent of Africa open to us. The friendly relations existing between these tribes and all Yorouba will give us in the near future the key to this immense country. Why may we not go very gradually on to this most desirable end?

We have commenced very cautiously the work of buying the slaves, particularly the children which we buy in the markets; we collect them on the large farms and accustom them to regular work while giving them instruction suitable to their position. In a few years these little slaves form free families, free with the true liberty of the children of God. The missionaries have to work slowly because they have only the offerings of the Society of the Propagation of the Faith to aid them instead of millions from the States or from humane societies. I have often been asked why Catholics and the real friends of civilization do not form a vast association for the special purpose of developing Christian influence, by establishing large agricultural colonies under the direction of the missionaries. This is an idea to which competent men could easily give a practical form, and I am convinced that in every instance the funds employed in this way would be returned to the association with interest after having aided in the regeneration of entire provinces.

North of Abeokouta the roads are always blockaded, and no one is allowed to pass, for the country is in a chronic state of war. Nearly all the men are under arms, and the tribes are dying of hunger notwithstanding the fertility of the soil. Not that they are killed in battle; but for years they have had to preserve an attitude of defence, always fearing an attack, and all this time the land has been left uncultivated. A chief said one day to one of our missionaries, "I will send you an ambassador, that you may tell him how to obtain the intervention of France to establish peace between the tribes, for we have been at war for years, and misery is at its height among us."

At the beginning of this year, Ogudipé succeeded in opening this northern route to two missionaries; he provided them with a guide, and they visited Yorouba. A great field is, I assure you, open to us here. There are cities as large as Abeokouta; and others smaller, but as thickly populated. Everywhere the population is dense and very accessible. The king of this country is powerful. He asks for missionaries. He sent an ambassador to Lagos to see how we are installed. The ambassador returned filled with admiration for all he had seen. In a few weeks three missionaries are going to fulfil the desire of the King of Oyo. One of the motives which impels the king to ask for the missionaries is to prevent Mahometanism from invading his people; it overruns and implants itself everywhere. "Men of God, come to our aid," cries the king.

Is it not a special dispensation of Divine Providence that this fetich-worshipping king should voluntarily call Catholic priests to resist Mussulman proselytism? This part of Africa no longer belongs to Mahomet, although he has already numerous disciples among its inhabitants. Islamism gradually spreads itself by successive encroachments. The Mussulmans are at first very benign, but as soon as they feel their numbers sufficient they impose their rule, and if necessary kill or sell as slaves those who resist, or at least force the natives to give up their lands and emigrate to other parts of the country, where they soon follow them and the same scenes are repeated. It is well known that the Mussulman rule is a combination of all the brutalizing influences of centuries. If, then, we love these poor blacks, who are our brothers, let us not lose time in coming to their aid. While they are fetichists, they are accessible and even desirous to receive Christian instruction. Let us then use every effort to bring to them that Christian civilization which we have always enjoyed, that liberty of the children of God which alone will make them perfect men.

Apostolic Prefecture of Dahomey.

The countries of which I have just spoken more than suffice to occupy all the energy of the chief of the missions. Moreover to the West is a region of country no less interesting and very thickly populated, but communication is very difficult, the language being different, and very urgent is the need of a missionary centre on this coast. Pope Leo XIII., by a brief of the 24th of June 1883, separated this portion of the territory from the Apostolic Vicariate of Benin, and made it the Apostolic Vicariate of Dahomey. This Prefecture extends along the coast from the Volta to the Okpara, and in the interior has no limits.

The only missions now in existence here are the station at Agoué and the school at Whyda.

Agoué is an agglomeration of some 8000 souls, governed, like all the neighboring towns, by a Cabécère of the King of Dahomey. Our establishment here dates from 1874. The missionaries have had to struggle against the profound ignorance and the disorders of some of the former slaves who returned from Brazil, and who adore nearly all the fetiches, while professing to be Christians. The missionaries, acting as schoolmasters, gradually effected a very sensible amelioration. A school for girls is established which gives us a sure element of regeneration for family life. Their congregation numbers nearly a thousand Christians, and no one even among the pagans seems to avoid them. The chief of war and the neighboring chiefs entrust their children to us to be educated. Two small pharmacies, one kept by the missionaries, the other by the Sisters, are fully appreciated by the blacks and give us an important influence among them.

The school at Whyda is kept by a catechist and his wife. The missionaries will soon establish themselves in this vicinity.

Four other schools are about to be established either at Popos or at Dahomey.

The interior is little known, but contains no doubt much that is interesting. We know, however, that there are there very populous cities. Salaga and Abomey have been visited by Europeans. The first contains more than 40,000 inhabitants, and is the centre of extensive commerce with Soudan. Abomey, the capital of the kingdom of Dahomey, has 30,000. Dâncbé, and other cities nearer the coast are also of importance. Atakpamé, which the blacks represent to us as having more than 30,000 inhabitants, seems to border upon the immense territory of the Manhis: this is one of the points which our missionaries have in view. These countries seem to have reached a time of transformation. Far beyond the limits where the direct influence of the missionaries and commercial relations have been exercised, a change is being produced in the minds of the people, of which the blacks themselves have often spoken. The gross superstitions of fetichism are fast falling into discredit, and the time is not far distant when Christianity will embrace all in this country who are not Mussulman.

Appendix.

Apostolic Prefecture of the Gold Coast and of the Ivory Coast.

There have been commercial houses on the Gold Coast for centuries, but there has never been till now a single Catholic missionary. The country is much more populous than Benin, but there was no trace of Catholicism to be found there. We established our first station at Elmina, a town of 15,000 or 18,000 souls. We could only begin with schools. That for boys already numbers nearly 2000 pupils. The girls' school was only established last March. The principal people with whom we have to deal in Elmina are the Fanti and the Achanti, among whom fetichism still reigns. We would have established ourselves this year in the capital of Achanti, but for the political revolution which deposed the king in 1882, at the time when our missionaries visited the capital. We have no reason to fear any opposition on the part of the new sovereign, but we prefer to wait till the city is in a more settled state before establishing ourselves there.

We have laid the foundation of another station at Axim, which will serve as an entering wedge to Vassaw and Apollonia.

We have only been two years and a half in charge of the Gold Coast. No idea can be had of the difficulties experienced by the missionaries at the outset, when establishing missions and schools with such very limited means.

The Ivory Coast, which forms part of this same mission, has not yet a single missionary. This coast is little frequented, but what we have learned of it excites very great interest. At various points merchants ship the Kroumans, who go in companies to work in the European factories. Merchants prefer them to all other blacks. The Kroumans are, in fact, industrious, diligent, and energetic. At home they devote themselves to agriculture, especially to the cultivation of rice. The greater part of them emigrate to different points on the coast just as soon as they have acquired a competency which will allow them to live quietly in their native country. Though they are much dreaded in war, they are a very peaceful people. Their religion is fetichism.

A mission among the Kroumans would do much good to these simple people, who are much less vicious than the other negroes. They are called the *Gauls of Africa*.

Apostolic Prefecture of the Niger.

A decree of the Propaganda, dated the 2d of May of this year (1884), has added to the territory already under our care that part of the interior of Africa which lies between the Niger and the Bénou. Two of our missionaries explored this country in 1883. The cities and towns are numerous and populous. Fetichism is the prevailing religion, but Mahometanism has already many adherents. In many places the inhabitants resist the invasion of the Mussulman. Our two explorers were well received by the kings and chiefs. Nearly everywhere a more or less exact report of the missionaries had preceded them, but their reputation was always that of men of God who do good and teach a good doctrine.

The French and English companies who have factories on the banks of the Niger and the Bénou promise us the co-operation of their agents to assist us in establishing ourselves. I do not know the character of these people, but the general idea we have of them leads us to hope that the labors of the missionaries among them will not be without rich results. Unfortunately, the few missionaries we are now able to scatter over such an immense space may not be able to effect such results as we would wish. Nevertheless, we must undertake the work courageously, leaving to God to fructify the little grain of mustard-seed sown by our efforts, and perhaps some day numerous birds of the air will dwell in the wide-spreading branches thereof.

Egypt.

Since the Crusades, Egypt has never been without missionaries. For a long time they were obliged to confine themselves to guarding the ancient sanctuaries and sustaining the faith in the souls of the faithful who surrounded them, which they often did at the price of their lives. It would take long to enumerate the martyrology of the children of St. Francis who in this glorious cause remained faithful unto death.

But even in this century Egypt has undergone many social transformations. Europe has exercised a salutary influence by her steady commercial relations. The cities on the coast, and Cairo itself, have important European colonies. The Christian Brothers and the Sisters of Charity are established here, and their schools contain large numbers of pupils. Nevertheless, those who know this country cannot but be deeply impressed with the great need it has of more missionaries. There are Franciscans in the cities on the coast, at Cairo, and in two other cities. But the Delta, with its population of Fellahin, so down-trodden, so overworked, and often so sadly oppressed by the extortions of pitiless masters, is entirely without Christian influence. The Holy See has authorized us to found stations among the Fellahin, and we cannot refuse to come to the aid of these thousands of unfortunate human beings. Moreover, His Holiness Leo XIII. has just erected this mission to an Apostolic Prefecture. We have opened schools for boys at Zagazig and at Tantah, but, having only the small offerings of the Society of the Propagation of the Faith with which to meet the expenses of rent and building, we have had to work slowly, and we have not been able to enlarge the schools.

The establishing of free schools will give us the greatest

and most unbounded influence in this country. Missionary labor in Egypt, just as among the pagans of Guinea, will have to be a work of patience and slow results. It is not the schools alone which are of immediate importance. It is little more than a year since we opened at Tantah a free dispensary for the sick. Sufferers of every kind crowd around the doors. There are only two Sisters to serve them, very few remedies, and often no linen to bind up their wounds. Even before the doors are open in the morning the people are waiting outside. The Sisters are obliged to close the doors in order to have time to take their meals, and even then the poor sufferers clamor at the windows. Two Sisters are utterly insufficient for this work, and those who have at heart the regeneration and salvation of these Fellahin, so worthy of interest, would do a good work by turning their attention to the city of Tantah, the most important in Egypt after Cairo and Alexandria. It is not a half-European city, but is composed entirely of Fellahin. It contains the tomb of the Marabut, called Saïd Ahmed of Bedaoin, the most venerated by Mahometans after that of Mahomet. The three great annual fairs attract here thousands of Arabs, among whom there are many maladies to be cured. A dispensary large enough and with sufficient means to admit of their being cared for in great numbers would exercise a most salutary influence and bring many blessings to the benefactors.

At Zagazig also we have opened a small dispensary. The people crowd here in such numbers that it is impossible to pass through the street, and the missionaries are obliged to have a strong negro to keep order, at least around the Sisters while they are dressing wounds. Each one cries out his malady and begs for a remedy. What a field is here open amidst all these human miseries for a skilful, zealous physician! Unfortunately, the missionaries have not the means to employ a physician, nor even to furnish anything like a complete pharmacy, or to build the hospital which is so indispensable. They do what

COLLEGE OF ST. LOUIS AT TANTAH (EGYPT) UNDER THE DIRECTION OF THE FATHERS OF THE SOCIETY OF AFRICAN MISSIONS.
(From a sketch by Rev. Fr. Ferdinand Merlini.)

they can, trusting to God to give to their simple remedies that perfect efficacy which will dispose the hearts of these poor people to receive their teachings.

The Fellahin are essentially agriculturists. We must follow them to their native plains, where they are most accessible, and bring to them the civilizing influence of the Christian faith. They hire themselves with their families as laborers. We must become farmers so as to employ these Fellahin and teach them to gain an honest livelihood. This is a project that we have long contemplated, and which I trust we are now about to realize. The great difficulty is to collect the funds necessary to purchase the land and make the first instalment. We would be at no expense for the Fellahin; they lodge themselves in their own inexpensive way.

To these agricultural colonies we intend to add agricultural schools, industrial schools, and orphanages.

The orphanages are always too small, so great is the number of vagrant children. The Egyptian Government is always very willing to surrender them to our care.

The industrial school is not for our pupils alone. The Christian Brothers have often urged us to establish this school for the benefit of their pupils also; for if the pupils on leaving their free schools are apprenticed to Mussulmans, Jews, Greeks, or freethinkers, these masters trouble themselves very little about the morals of their apprentices unless to destroy them, which is often the case.

The agricultural school is a practical speculation. The chiefs of the villages willingly send their children to learn a process of cultivation which will make their rich lands capable of much more abundant harvests. We do not wish to effect a great transformation by supplanting manual labor by machinery, but only to render labor more fruitful by making it more intelligent.

We also teach these poor Fellahin how to secure for themselves an honest division of their crops without having recourse to the usurers established in all the villages, who make them pay a commission of fifty or sixty per cent, and sometimes even more. They pledge their crops for one or two years, leaving nothing for themselves. The poor Fellahin works on from year to year, the usurer taking every season the fruits of his labor, and the Fellahin continues to contract new debts and new obligations.

We do not deem it unworthy of the missionaries to procure for God's oppressed creatures that alleviation which, though only temporal, makes them understand that the missionaries are their friends. They listen willingly to those whom they know to be friendly; and although we do not expect to make any immediate conversions, we have every reason to hope that our efforts will bring these poor people, so long subject to the yoke of Mahomet, nearer to God, and that their children, enlightened by Divine Grace, will flock in crowds to the fold of the Divine Shepherd.

This, my friends, is a brief sketch of the effects of Catholic charity in that part of the Lord's vineyard which has been entrusted to the care of our Society. The results have been obtained at the price of great sacrifices. We must now make these results permanent and extend them. Pray, my friends, that the Master of the harvest may send many laborers, and that generous souls may be inspired to furnish them with the material resources necessary for the work.

SPIRITUAL PRIVILEGES GRANTED TO THE ASSOCIATES OF THE AFRICAN MISSIONS.

PLENARY INDULGENCES.

1st. On the day of reception.

2d. On the Feast of the Crown of Thorns.

3d. On the Feast of the Exaltation of the Holy Cross.

4th. At the hour of death, on invoking the Holy Name of Jesus.

An indulgence of sixty days each time an Associate performs a good work for the success of the African Missions.   Exequatur,

✠ CARD. J. J. M. DE BONALD,
Archbishop of Lyons.

CONDITIONS OF ADMISSION TO THE ASSOCIATION OF THE AFRICAN MISSIONS.

1st. Those are *Affiliated* who give $1 00
2d. Those are *Protectors* who give annually..... 125 00
3d. Those are *Founders* who give................. 3,000 00

This last amount forms a fund for the perpetual maintenance of a missionary in the Society.

I. A Mass will be offered at a privileged altar every Friday of the year by the Superior-General of the Society for the Associates.

II. Twenty Masses will be annually offered for each Protector.

III. An *annuel* of 365 Masses will be offered for each Founder.

Offerings will be received by Rev. Fathers Merlini and Connaughton, who have charge of the collection at the House of the Immaculate Virgin, care of Rev. Father Drumgoole, Lafayette Place, New York City, P. O. Box, 3512.

Offerings may also be sent to the Superior-General of the Society of African Missions, Cours Gambetta 174, Lyons, France.

Approbations.

APPROBATIONS OF THE MOST REV. AND RIGHT REV PRELATES OF AMERICA.

The Rev. Fathers Desribes and Merlini, of the Society of African Missions, having been duly authorized by his Eminence, the Cardinal Prefect of the Propaganda to come to this country for the purpose of collecting funds to assist them in sustaining and extending the highly important and meritorious work which has been confided to them are hereby permitted to solicit contributions from the faithful of this diocese, to whose charity they are commended.

NEW YORK, January 7, 1881.
✠ JOHN CARD. MCCLOSKEY,
Archbishop of New York.

The Fathers Desribes and Merlini have permission to apply to the reverend clergy of the diocese for assistance in their collection among the faithful for their missions of Africa.

BOSTON, Feb. 22d, 1881.
✠ J. J. WILLIAMS,
Archbishop of Boston.

The Rev. Fathers Merlini and Gallen have permission to solicit aid in the archdiocese of Chicago in behalf of their African missions, with the consent of the reverend clergy.

CHICAGO, May 31, 1881.
✠P. A. FEEHAN,
Archbishop of Chicago.

Some time after the coming spring, Father Merlini Apostolic Missionary, may make an appeal to any church of the diocese, to which he will be invited by the pastors in charge.

BALTIMORE, December, 1884.
✠ JAMES GIBBONS,
Archbishop of Baltimore.

Approbations.

I recommend the appeal of the Reverend Fathers of the African missions to the Catholics of St. Louis.

BALTIMORE, November 28, 1884.
✠ PETER RICHARD KENRICK,
Archbishop of St. Louis.

I authorize the Fathers of the African missions to collect in the diocese of Cincinnati, with the consent of the reverend pastors, and I warmly recommend this work of charity.

BALTIMORE, November 28, 1884.
✠ WILLIAM HENRY ELDER,
Archbishop of Cincinnati.

The following Bishops have granted the same permission:

✠ Right Rev. Francis McNeirny, Bishop of Albany.
✠ Right Rev. P. T. O'Reilly, Bishop of Springfield.
✠ Right Rev. T. F. Hendricken, Bishop of Providence.
✠ Right Rev. J. A. Healy, Bishop of Portland.
✠ Right Rev. D. M. Bradley, Bishop of Manchester.
✠ Right Rev. M. J. O'Farrell, Bishop of Trenton.
✠ Right Rev. F. S. Chatard, Bishop of Vincennes.
✠ Right Rev. J. O'Connor, Vicar Apostolic of Nebraska.
✠ Right Rev. J. A. Watterson, Bishop of Columbus.
✠ Right Rev. J. Ireland, Bishop of St. Paul.
✠ Right Rev. J. Hennessy, Bishop of Dubuque.
✠ Right Rev. W. M. Wigger, Bishop of Newark.
✠ Right Rev. John Loughlin, Bishop of Brooklyn.
✠ Right Rev. John Lancaster Spalding, Bishop of Peoria.

www.ingramcontent.com/pod-product-compliance
Lightning Source LLC
Chambersburg PA
CBHW031618170426
43195CB00037B/1129